D1376521

DETERMINATION OF ORGANIC REACTION MECHANISMS

DETERMINATION OF ORGANIC REACTION MECHANISMS

BARRY K. CARPENTER

Department of Chemistry
Cornell University
Ithaca, New York

A Wiley-Interscience Publication

John Wiley & Sons

New York • Chichester • Brisbane • Toronto • Singapore

Library of Congress Cataloging in Publication Data:

Carpenter, Barry K. (Barry Keith), 1949–
   Determination of organic reaction mechanisms.

   "A Wiley-Interscience publication."
   Includes bibliographical references and index.
   1. Chemistry, Physical organic.   I. Title.
QD476.C37   1984   547.1'39   83-16845
ISBN 0-471-89369-2

Printed in the United States of America

10 9 8 7 6 5 4 3 2 1

To Joyce

PREFACE

This book is intended primarily for graduate students in organic chemistry although it might be of interest to professional chemists engaged in organic research. Its purpose is to familiarize the reader with the techniques available for studying reaction mechanisms in organic chemistry. Several aspects of the problem are considered, beginning with abstract philosophical concepts and finishing with a discussion of specific mechanistic investigations. It is hoped that it will provide the reader with a sufficiently solid foundation to begin his or her own research in the area.

The book is based on a one-semester course given by the author to first-year graduate students in the Department of Chemistry at Cornell University. It is intended for students who have already had substantial exposure to organic chemistry, probably including a graduate-level course. The reader is also assumed to have familiarity with the common techniques of organic spectroscopy such as infrared, Raman, and nuclear magnetic resonance spectroscopy. Previous exposure to kinetics and thermodynamics is assumed although elementary topics in these areas are briefly reviewed. Some familiarity with elementary linear algebra would be helpful to the reader but this is not essential.

The techniques considered are: isotopic labeling, chirality and stereochemistry, kinetics, isotope effects, methods in acid–base chemistry, interpretation of activation parameters, and direct detection of reactive intermediates. The section on interpretation of activation parameters includes discussions on group additivity calculations, linear free energy relationships such as the Hammett, Brønsted, and Grunwald–Winstein equations, and a short section on analysis of activation volumes. The kinetics section presents a convenient way of handling the kinetics of unimolecular arrays through the use of linear algebra. Each technique is discussed in terms of its theoretical background, where appropriate, and its use is then illustrated with at least one example from the research literature.

The final chapter presents a detailed analysis of specific case histories of mechanistic investigation. This chapter serves to integrate all of the previous material. In addition it shows how the desire for "elegance" and rigor emphasized in the previous pages must be tempered with the needs of practicality and with the finite nature of research funds.

BARRY K. CARPENTER

Ithaca, New York
December 1983

CONTENTS

DETERMINATION
OF ORGANIC
REACTION MECHANISMS

CHAPTER **1**

PHILOSOPHY

It will be clear to many readers that the title of this book is, strictly speaking, inaccurate. One can never *determine* the mechanism of any reaction with 100% certainty. This limitation is a direct consequence of the fact that chemistry is an empirical science. Other such limitations and a strategy to minimize their effect will be the topics of this chapter.

The intention is not to teach the reader any philosophy, since the author is not qualified to do so; rather it is to lay out the foundation for all the subsequent chapters as clearly and explicitly as possible. In particular the strategies of how and when to do particular experiments will be governed in part by the considerations discussed in this chapter. Readers who are interested in a more detailed and authoritative discussion of the philosophy of science are recommended to read the books by Popper[1] and Hempel.[2]

The concepts of empirical and nonempirical science are intimately involved with the concepts of inductive and deductive logic. An illustration will perhaps serve to clarify the connection.

Let us treat the proposition that the internal angles of a plane triangle sum to 180° as a hypothesis to be tested by the methods of empirical science. The procedure would be to draw some number of triangles, say 20, and then to measure and sum the internal angles for each. A typical result might be

Sum = 179.7 ± 0.4°   (standard deviation, 20 readings)

Should one consider such a result as support for or disproof of the hypothesis? The probability that the true sum is 180° is high but not 100%. The probability that the true sum is 179.7° is actually somewhat higher. Furthermore, the probability that the true sum is 200° is low but not 0%. Immediately one can see that the comparison of experimental results with theoretical hypotheses is going to

1

be a difficult problem that will have to be couched in statistical terms. But there is another problem. In formulating this thought experiment we have implicitly assumed that all plane triangles have the same value for the sum of their internal angles. Unfortunately the methods of empirical science can never prove that this is true. One cannot, by this procedure, be certain that there is not somewhere a very special triangle whose internal angles sum to, say, 110°.

Despite these uncertainties, most people would be willing to accept that experimental data such as those presented above did support the original hypothesis. This step from the specific (measurement on 20 triangles) to the general (statement about triangles) is called inductive logic, although one can recognize that it is really more an act of faith than an exercise of true logic. It is the frequent failure of such extrapolations that leads to the repeated revision of scientific theories, thereby keeping many of us in business!

In the particular example of the triangle problem there is, of course, another way to test the original hypothesis. If one accepts two axioms, namely, that the angle described by three colinear points is 180° and that the interior alternate angles between parallel lines are equal, then it is possible to use the theorems of plane geometry to *prove* that the sum of the internal angles of *any* triangle is *exactly* 180°. This is the method of nonempirical science. It allows one to use deductive logic to go from the general statement about all triangles to a specific statement about any particular triangle of interest.

Unfortunately, empirical sciences such as chemistry have no theorems and so one must make do with the imperfect approach of inductive logic. The challenge, then, is to construct a protocol that will give one the greatest possible confidence in the general conclusions that are drawn as the result of some finite set of experiments. Probably the best procedure yet devised is an iterative cycle:

1. Formulate a hypothesis to fit the known facts.
2. Design and execute an experiment to test the hypothesis.
3. If the experimental results are consistent with the hypothesis (within the limits of experimental error) go to step 4, otherwise go to step 1.
4. If "all" of the testable features of the hypothesis have been subjected to experimental scrutiny then stop, otherwise go to step 2.

Note that one starts this cycle with the formulation of a hypothesis. It is a common misconception, nicely described by Hempel,[2] that a truly objective scientist gathers all of the relevant facts without prejudice *before* formulating any theory. A moment's thought will show that this is impossible. One cannot know which facts are relevant unless one has some hypothesis in mind. The task of gathering facts would go on forever! The objectivity comes not in the collection of facts but in their interpretation. One should be willing to admit that the experimental facts show one's pet theory to be incorrect (although when a reputation or career is at stake this might not be so easy![3]).

The word "all" in step 4 of the cycle presents some problems. The experimenter must be able to see what experimentally testable predictions are made by the hypothesis under consideration. Any one that is missed could be the fatal flaw that will destroy the whole theory at some future date. On a more practical note, the process of testing "all" of the predictions of some hypothesis can be tedious and time consuming. For this reason the exit from step 4 tends to occur when one has gathered enough data for a paper or thesis, when the funds for the research run out, or when the investigator simply loses interest in the problem.

Two important and uncomfortable conclusions emerge from this analysis:

1. One can never get closer to the truth than one's best guess.

2. That guess can never be proven correct. It can only be proven incorrect.

It becomes clear, then, that the process of formulating a hypothesis is all-important. Regrettably (or not, depending on one's point of view) there are no algorithms for generating scientific hypotheses. There are, however, some criteria that ensure that any hypothesis meets certain minimum standards:

(a) It must be consistent with all of the available experimental data. It can be neutral about some facts but should not be in direct conflict with any unless there is very good *independent* evidence to suggest that the data are faulty.

(b) It must make experimentally testable predictions that would, if not verified, be capable of proving it false.

(c) Entia non sunt multiplicanda proctor necessitatum (Occam's razor). In the event that several hypotheses are found to fit the facts, the simplest is given preference. Note that a hypothesis might be more complex than one of its rivals for a particular example but still be preferable if it encompasses more cases.

(d) Where possible ad hoc additions to a hypothesis as devices to explain away inconsistencies with experimental facts should be avoided. A clear example is the negative mass that had to be attributed to phlogiston when it was discovered that substances increased in mass upon combustion. Unfortunately not all cases are this clear-cut. The dividing line between "refinement" of a model and ad hoc "patching up" to make it fit the facts is often hard to identify. Fortunately hypotheses with an inordinate number of "patches" tend to become susceptible to replacement by operation of Occam's razor.

In this book we will be concerned with the formulation and testing of hypotheses about reaction mechanisms. These hypotheses are subject to additional constraints that are of special concern to the physical sciences, namely the laws of thermodynamics, symmetry, statistics, and mass conservation. Even these "laws" are themselves unproven hypotheses but they are so thoroughly rooted at the foundations of the physical sciences that they have come to occupy positions similar to the axioms of mathematics. A mechanism that violated the second law of thermodynamics would require a mountain of supporting experimental data before it would be generally accepted! On the other hand, one need not feel so constrained by "rules" that are based on simplified models of chemical behavior. Thus a reaction mechanism that appeared to violate the Woodward–Hoffmann rules of orbital symmetry conservation could be acceptable and might even be correct, although it would probably be desirable for the proponent to explain why the rules were not followed in this case.

REFERENCES

1. K. R. Popper, *The Logic of Scientific Investigation*, Basic Books, New York, 1959.
2. C. Hempel, *The Philosophy of Natural Science*, Prentice-Hall, Englewood Cliffs, N.J., 1973.
3. R. Estling, *New Scientist*, **96,** 808 (1982).

CHAPTER 2

ISOTOPIC LABELING

In its simplest form the isotopic labeling experiment merely provides a way of keeping track of where the atoms of a reactant end up in the final product. There are more sophisticated versions of the technique that we will see later in the chapter.

A convenient way of determining whether isotopic labeling is appropriate for solving a particular mechanistic problem is to number all of the atoms of the reactant and then to determine whether the various mechanisms under consideration lead to different connectivities in the isolable product(s). If they do not, no labeling experiment will solve the problem. If they do then there is *in principle* a labeling experiment that will work. In practice one might be limited by difficulties of synthesis or by the number of isotopes available for a particular element.

Figure 2.1 illustrates the numbering test for the conversion of cyclododeca-1,5,9-triyne to hexaradialene. This example shows that mechanisms A and B could be distinguished (in principle) by a labeling experiment, whereas no such experiment would serve to distinguish between mechanisms B and C.

In the sections that follow we will see the types of mechanistic information that can be gained from labeling experiments. Discussion of the kinetic and thermodynamic changes that often accompany the introduction of isotopic labels into a molecule will be deferred until Chapter 5.

2.1. SYMMETRIES OF INTERMEDIATES

The mechanism of a reaction is not completely defined until one has determined the structures of any reactive intermediates. Frequently these intermediates are not observable under the reaction conditions and so their structures must be

5

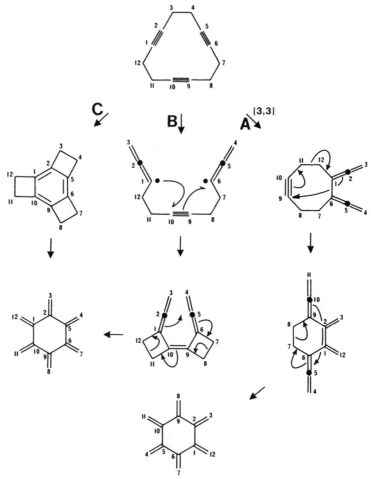

FIGURE 2.1. Test of applicability of isotopic labeling to the rearrangement of cyclododeca-1,5,9-triyne. In mechanism B the initial cleavage could have involved the 7,8 or 11,12 bonds, but the final product would have been the same. In mechanism A it is unnecessary to consider the other two possible [3,3] shifts of the reactant since the one depicted already gives a different product from B or C.

deduced indirectly. Labeling experiments can be useful when a potential inter-mediate has higher symmetry than the reactant that is its precursor. A classic example of this kind of labeling experiment is the work of Roberts and cowork-ers[1] on benzyne.

Chlorobenzene reacts with potassium amide in liquid ammonia to afford aniline:

Perhaps the simplest mechanism for this reaction would be a nucleophilic aromatic substitution (A). However, the products derived from substituted chlorobenzenes show that something else must be going on. The observation of a *meta*-product might have meant that some other process was occurring in parallel with mechanism A, but this explanation was rendered untenable by an isotopic labeling experiment. It would be unacceptably ad hoc to claim that mechanism A and the other process just happened to occur at the same rate. This result illustrates clearly the advantage of using isotopic labels over "chemical" labels such as methyl groups. Provided there are no large isotope effects, one can often apply simple statistics to predict ratios of label-isomeric* products for a given mechanism or, working in reverse, can deduce something about the structures of intermediates by determining these ratios experimentally. In the present example the 1:1 ratio of label-isomeric anilines strongly implied the existence of an intermediate in which the erstwhile *ipso-* and one *ortho*-carbon had become equivalent by symmetry. Plausible candidates for this intermediate were benzyne (mechanism B) and a bicyclic aziridine (mechanism C). Distinction between these two mechanisms—with the results favoring the benzyne

---

*Label isomers are molecules that differ only in the location of an isotopic label.

intermediate—was achieved by Roberts' group using a mixture of labeling and isotope effect studies.[1]

A somewhat different example of the use of isotopic labels to deduce symmetries of intermediates is provided by the work of Whitman and Carpenter[2] on cyclobutadiene. Here the choice was between an intermediate with twofold symmetry—rectangular cyclobutadiene—and one with fourfold symmetry—square cyclobutadiene. Earlier experimental studies had concentrated on infrared spectroscopy of cyclobutadiene prepared in a solid argon matrix (see Chapter 8) but Whitman's experiment involved solution phase chemistry of the compound. The key to this experiment is to note that vicinal double labeling of cyclobutadiene results in two label isomers for the rectangular structure but only one for the square:

E = COOCH₃

Whitman was able to show that the azo-compound **1**, when deazetized to produce cyclobutadiene-d_2 in the presence of a high concentration of methyl (Z)-3-cyanoacrylate, afforded products **2a–d**, and that $[2a] + [2b] \gg [2c] + [2d]$. This result is inconsistent with a square structure for cyclobutadiene since the square should have given essentially equal proportions of all four label-isomeric products. Again, the use of isotopic labels enabled the authors to make experimentally testable predictions based on simple statistics. Later work by Whitman[3] allowed limits to be placed on the activation parameters for interconversion of the two rectangular isomers by studying the concentration and temperature dependence of the product ratios.

2.2. INTER- VS. INTRAMOLECULAR REACTIONS

Unimolecular rearrangements are usually considered to proceed either by intramolecular or fragmentation–recombination (intermolecular) mechanisms. Some version of the double-labeling crossover experiment generally serves to distinguish between the two.

The idea behind this type of experiment is to doubly label the reactant molecule in such a way that one isotopic label would be in each of the fragments for the hypothetical intermolecular mechanism. This doubly labeled reactant is then mixed with an equal quantity of unlabeled reactant and the mixture submitted to the reaction conditions. The expected results are shown below for the generalized rearrangement $X—Y—Z \longrightarrow Y—Z—X$:

A: $X—Y—Z + X^*—Y^*—Z \longrightarrow Y—Z—X + Y^*—Z—X^*$

B: $X—Y—Z + X^*—Y^*—Z \longrightarrow X + X^* + Y—Z + Y^*—Z$
   $\longrightarrow Y—Z—X + Y^*—Z—X^* + Y—Z—X^* + Y^*—Z—X$

The intramolecular mechanism (A) affords only unlabeled and doubly labeled products whereas the intermolecular mechanism (B) leads to the formation of singly labeled products by recombination of fragments from different reactant molecules. The detection of singly labeled products in such an experiment is unambiguous proof of an intermolecular component to the mechanism. However, the formation of such products in an intermolecular process can occur only if the fragments are present in a common pool. Thus a fragmentation–recombination reaction that occurs within a "solvent cage" cannot be distinguished from an intramolecular rearrangement by a crossover experiment. Indeed, it is frequently difficult to find any kind of experiment that will allow one to make this type of distinction.

A minor variation on the double-labeling crossover experiment is to use two singly labeled reactants such as $X^*—Y—Z$ and $X—Y^*—Z$ and then to look for doubly labeled and unlabeled products as evidence for an intermolecular mechanism. The choice between the two versions of the experiment usually depends on the ease of synthesis of the labeled reactants.

There is a more significant variation that can be advantageous if one can be sure of identifying the precise nature of the potential fragments in the intermolecular mechanism, and if one of these fragments is available as a reagent. Under these circumstances one can add the fragment in isotopically labeled form to the reaction mixture and look for incorporation of the label into the final product(s). This technique usually presents a less formidable synthetic challenge but provides more restricted information in that dissociation of the reactant into fragments other than those assumed by the experimenter will not be detectable. The work of Grovenstein[4] on anionic 1,2 rearrangements provides an interesting example of the incorporation technique:

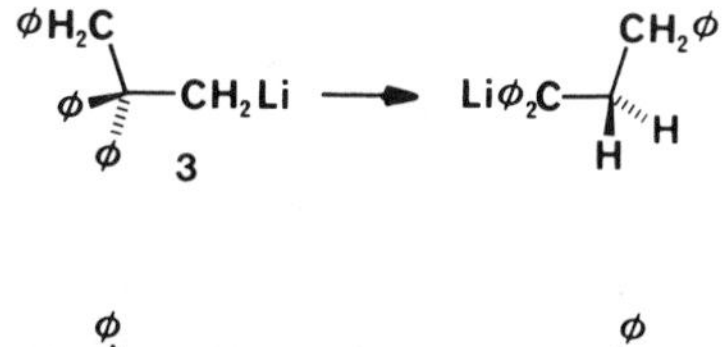

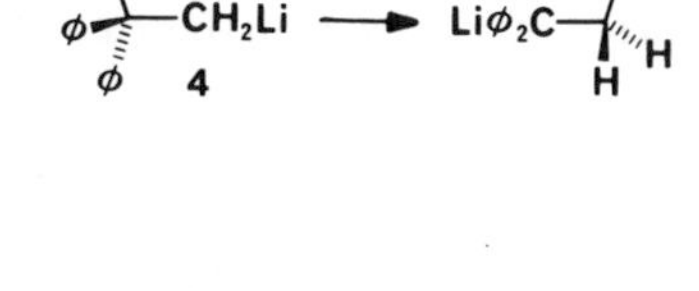

The alkyl lithium compounds **3** and **4** were found to undergo 1,2 rearrangements that appeared to be analogous. However, addition of ^{14}C-labeled benzyl lithium to **3** resulted in incorporation of the label into the final product whereas addition of ^{14}C-labeled phenyl lithium to **4** led to no incorporation.

These results are consistent with an intermolecular mechanism for the rearrangement of **3**—involving fragmentation to 1,1-diphenylethylene and benzyl lithium—but an intramolecular mechanism for the rearrangement of **4**. The reasons for the difference in mechanism are of interest but not relevant to the present discussion.

The work of Curtin on ester pyrolysis[3] illustrates an alternative solution to the problem of distinguishing inter- from intramolecular reactions. The ester **5** affords *trans*-stilbene and benzoic acid on pyrolysis at 400°C. One could imagine an intramolecular ($_\sigma 2_s + _\sigma 2_s + _\pi 2_s$) mechanism (A) or a dissociation–disproportionation mechanism (B). At first sight this looks like another problem to be solved by the double-labeling crossover technique. In principle the experiment could be performed by labeling the benzoyl moiety and the methylene hydrogens of the reactant and then looking for singly labeled benzoic acid in the products from pyrolysis of doubly labeled and unlabeled esters. In practice the experiment would be thwarted by the facile exchange of carboxyl hydrogens between benzoic acid molecules—a process that would ensure the formation of singly labeled products regardless of the mechanism of the pyrolysis reaction.

Curtin's solution to the problem was based on the recognition that the methylene hydrogens are diastereotopic in **5** and would remain so throughout mechanism A. Mechanism B, on the other hand, would produce a radical (**6**) in which the methylene hydrogens were enantiotopic. In more practical terms, the methylene hydrogens should remain chemically distinguishable by mechanism A but should become chemically equivalent by mechanism B.

Accordingly, Curtin and coworkers prepared the *threo* and *erythro* label isomers **7t** and **7e**. The expected products from these labeled reactants are shown for each mechanism in Figure 2.2. The crucial feature that would serve to distinguish between the two mechanisms is that **7t** and **7e** should give the same ratio of deuterated:undeuterated stilbene by mechanism B but different ratios by mechanism A. If one assumes that the intramolecular mechanism would adopt a chairlike transition state and that the phenyl groups would prefer to occupy pseudoequatorial positions then one can extend the prediction to say that mechanism A should give mostly unlabeled stilbene from **7t** and mostly labeled stilbene from **7e**.

Experimentally Curtin and coworkers found that **7t** gave 82% stilbene-d_0 and 18% stilbene-d_1 whereas **7e** gave 1% stilbene-d_0 and 99% stilbene-d_1. This result is consistent with mechanism A but inconsistent with mechanism B. The greater selectivity in the pyrolysis of **7e** is presumably the result of a hydrogen/deuterium primary isotope effect (see Chapter 5).

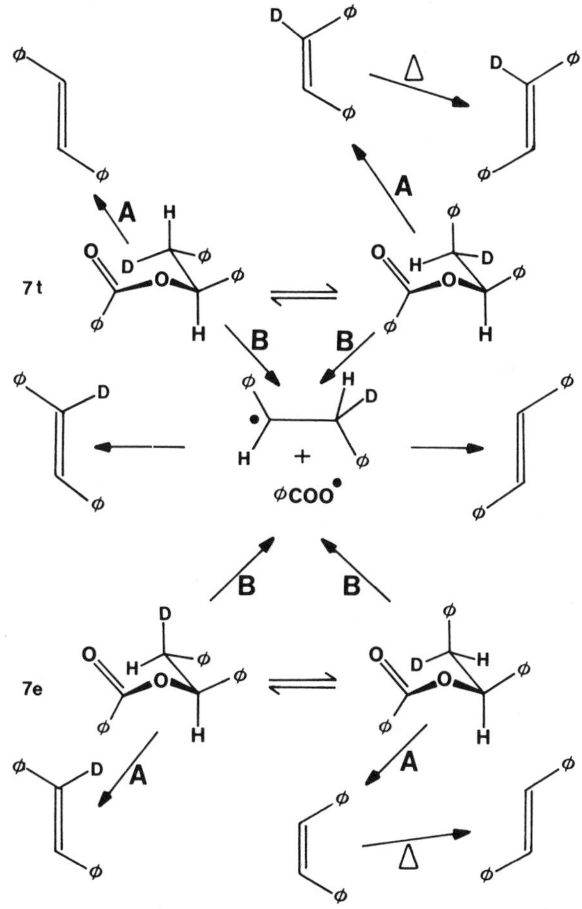

FIGURE 2.2. Expected products from the concerted (A) and fragmentation/disproportionation (B) mechanisms for the elimination of benzoic acid from the two diastereomers of 1-benzoyl-1,2-diphenylethane.

REFERENCES

1. J. D. Roberts, H. E. Simmons, Jr., L. A. Carlsmith, and C. W. Vaughan, *J. Am. Chem. Soc.,* **75,** 3290 (1953); J. D. Roberts, D. A. Semenow, H. E. Simmons, Jr., and L. A. Carlsmith, *Ibid.,* **78,** 601 (1956); J. D. Roberts, C. W. Vaughan, L. A. Carlsmith, and D. A. Semenow, *Ibid.,* **78,** 611 (1956).
2. D. W. Whitman and B. K. Carpenter, *J. Am. Chem. Soc.,* **102,** 4272 (1980).
3. D. W. Whitman and B. K. Carpenter, *J. Am. Chem. Soc.,* **104,** 6473 (1982).
4. E. Grovenstein and G. Wentworth, *J. Am. Chem. Soc.,* **89,** 1852, 2348 (1967).
5. D. Y. Curtin and D. B. Kellom, *J. Am. Chem. Soc.,* **75,** 6011 (1953).

CHAPTER 3

CHIRALITY AND STEREOCHEMISTRY

In this chapter we will consider the mechanistic information that can be determined by studying configuration changes at a chiral center.

3.1. RACEMIZATION STUDIES

The simplest experiments involving optical activity measurements are those in which one attempts to determine whether the products derived from an optically active reactant are racemic or are themselves optically active. This type of experiment is usually undertaken when one of the possible mechanisms under consideration would involve the formation of an achiral intermediate, thereby mandating the formation of racemic products. Should the products from such an experiment turn out to be optically active, it is then sometimes desirable to determine their optical purities in order to find out whether partial racemization could have occurred.

A nice example of this kind of study is provided by the work of Schultz[1] on the photorearrangement of **1** to **2**. A plausible precursor to **2** is the spirocyclohexadienone **3**, which could undergo a thermal hydrogen migration to give the observed product. Two mechanisms for formation of **3** might be the photochemical [1,3]sigmatropic shift A or the biradical route B. The crucial observation is that the biradical in mechanism B possesses a plane of symmetry and is therefore achiral. Consequently mechanism B would necessarily result in the formation of racemic **2** from optically active **1**. Mechanism A, on the other hand, could give a racemic product only by an improbable coincidence.

Schultz found experimentally that optically active **1** gave optically active **2**, a result that rules out mechanism B as the exclusive pathway for the rearrangement. Further information could be obtained by determining the optical purities of reactant and product. [Optical purity is defined as percent enantiomeric excess: the optical purity of an $x:1$ ratio of enantiomers is $|100(x-1)/(x+1)|$.]

The optical purities were determined by using the optically active NMR shift reagent Eu(hfc)$_3$, which is tris-[3-(heptafluoropropylhydroxymethylene)-d-camphorato] europium(III). Optically active shift reagents of this kind work by converting a mixture of enantiomers into a mixture of diastereomeric complexes that usually (but not always) exhibit different chemical shifts in the NMR spectrum. The enantiomer ratio of the original mixture can then be determined by integration of the NMR resonances.

In the present example the reactant was found to be 78.5% optically pure by this method. Unfortunately the product did not show resolved resonances upon treatment with Eu(hfc)$_3$ and so it had to be further derivatized by treatment with phenyl selenyl chloride. The resulting phenyl selenide was found to have an optical purity of 77.8%. Thus one can say that less than 1% of the photochemical rearrangement could have followed mechanism B. In fact it is quite possible that a small amount of racemization could have occurred during the derivatization of the product and that none of the rearrangement followed the biradical pathway.

In order to see a somewhat more complex example of a racemization study we turn to some work by Jensen[2] on the electrophilic cleavage of mercury carbon bonds.

Dialkyl mercury compounds undergo a conproportionation reaction with mercuric bromide to give two equivalents of an alkyl mercuric bromide:

$$R\text{---}Hg\text{---}R + Br\text{---}Hg\text{---}Br \longrightarrow 2Br\text{---}Hg\text{---}R$$

Jensen wished to determine whether this process occurred with retention, inversion, or racemization at the transferred alkyl group.

In order to do so he first resolved a sample of *sec*-butyl mercuric bromide (**4**) by displacement of the bromine with optically active mandelic acid salt. This was followed by separation of the resulting diastereomers and then reconversion to the bromide by treatment with HBr. The optically active *sec*-butyl mercuric bromide was then treated with *sec*-butyl magnesium bromide (which cannot be resolved) to give one optically active and one *meso* diastereomer of di(*sec*-butyl) mercury (**5**).

The expected outcome for the three mechanisms of the reaction of 5 with mercuric bromide can be summarized as follows:

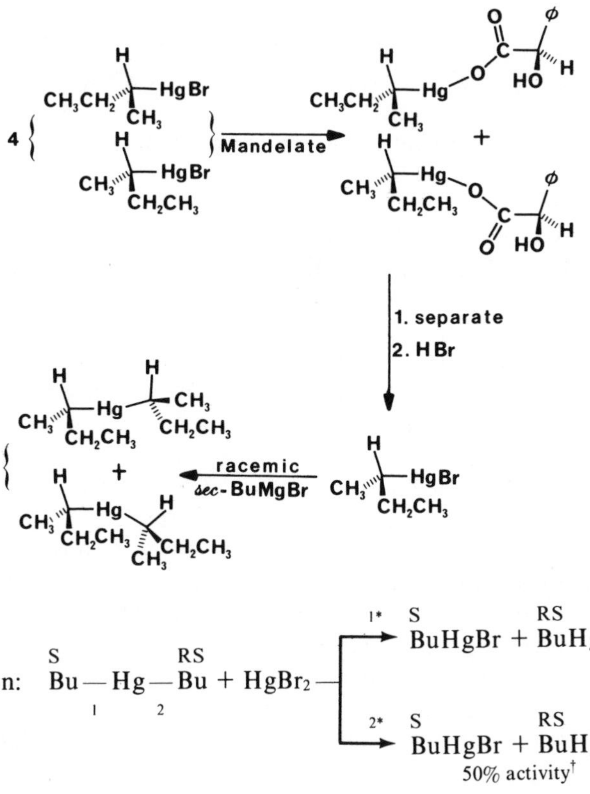

1. separate
2. HBr

racemic
sec-BuMgBr

Retention:

$$\underset{1}{\overset{S}{Bu}} - Hg - \underset{2}{\overset{RS}{Bu}} + HgBr_2$$

1* → $\overset{S}{Bu}HgBr + \overset{RS}{Bu}HgBr$

2* → $\overset{S}{Bu}HgBr + \overset{RS}{Bu}HgBr$
50% activity†

Inversion:

$$\underset{1}{\overset{S}{Bu}} - Hg - \underset{2}{\overset{RS}{Bu}} + HgBr_2$$

1* → $\overset{R}{Bu}HgBr + \overset{RS}{Bu}HgBr$

2* → $\overset{S}{Bu}HgBr + \overset{RS}{Bu}HgBr$
Racemic

*This number refers to that of the Hg—C bond cleaved.

†Optical purity of product with respect to 100% optically pure *sec*-butyl mercuric bromide used in the synthesis of **5**.

$$\text{Racemization:} \quad \underset{1}{\overset{S}{Bu}} - Hg - \underset{2}{\overset{RS}{Bu}} + HgBr_2$$

$$\xrightarrow{1^*} \underset{}{\overset{RS}{BuHgBr}} + \underset{}{\overset{RS}{BuHgBr}}$$

$$\xrightarrow{2^*} \underset{}{\overset{S}{BuHgBr}} + \underset{}{\overset{RS}{BuHgBr}}$$
$$25\% \text{ activity}^\dagger$$

This analysis assumes that the two diastereomers comprising the mixture **5** are formed in equal amounts in the synthesis and that they react at the same rate with mercuric bromide. Both assumptions seem reasonable since the distance between the two chiral centers is probably too large for their mutual interaction to be significant.

Jensen's experimental result was that the reaction product exhibited 48.5 ± 2.1% optical activity with respect to that of the resolved *sec*-butyl mercuric bromide used in the synthesis. Thus it appears that the electrophilic cleavage reaction occurred with retention of configuration at the alkyl group.

3.2. CORRELATION OF REACTANT AND PRODUCT CONFIGURATIONS

For some mechanistic problems it is not sufficient merely to determine optical purities of products, rather it becomes necessary to correlate the configurations of the reactant and product(s). This invariably represents a significant increase in the amount of experimental work. Indeed, the process of correlation is often the most time-consuming part of such an experiment.

Two approaches are available for the task of correlation. One is to determine the absolute configurations of the reactant and product(s), usually by chemical conversion to (or synthesis from) a natural product of known absolute configuration. The other is to interconvert the reactant and product by a sequence of reactions whose stereochemistries have been established. A variation on the latter involves converting the reactant and product to a common derivative, again by reactions of known stereochemistry. We will see examples of both general approaches throughout this book.

The elegant study of the semipinacol rearrangement by Collins and coworkers[3] is illustrative of the type of information available from experiments of this kind.

The amine **6** affords ketone **7** upon diazotization with nitrous acid. The first intermediate is almost certainly a diazonium ion but then the question is whether this forms a discrete carbonium ion (mechanism A) or rearranges in concert with the nitrogen expulsion (mechanism B). The carbonium ion of mechanism A would be achiral provided it could undergo internal rotations at a rate faster than the rate of rearrangement. The obvious first experiment is thus a racemization study. Accordingly, Collins and coworkers resolved the amine **6** with camphorsulfonic acid, obtaining the (+) enantiomer of unknown absolute configuration. Optically pure (+) **6** gave **7** with a specific rotation of +158.5°. This result showed immediately that passage through a rotationally equilibrated carbonium ion could not be the exclusive mechanism for the rearrangement. Repeated recrystallization of the product **7** increased its specific rotation to +210°, showing that the initial reaction product had been $100 \times 158.5/210 = 75.5\%$ optically pure. In other words, the reaction had proceeded either with 87.8% retention and 12.2% inversion or *vice versa*. To find out which, it was necessary to correlate the configurations of **6** and **7**. This was achieved by the sequence shown in Figure 3.1.

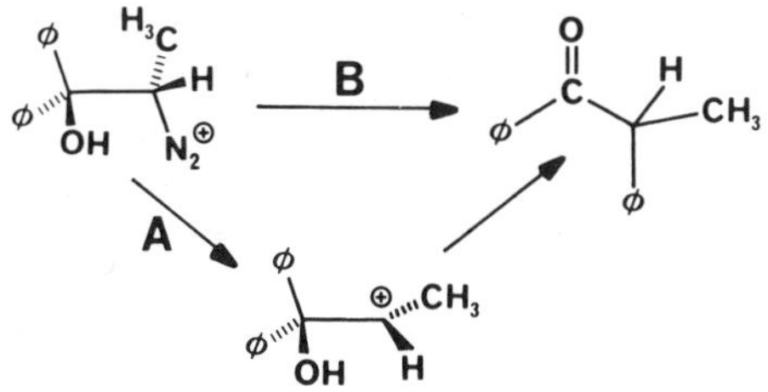

It was not necessary to know the absolute configuration of the (+) phenyl-

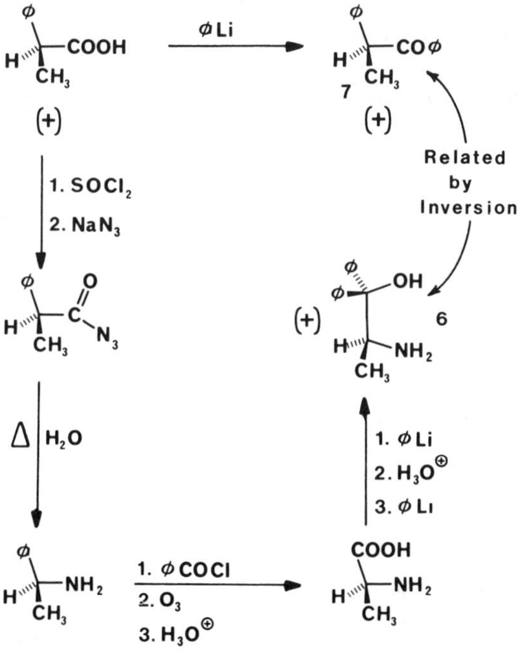

FIGURE 3.1. Correlation of reactant (**6**) and product (**7**) configurations for the semipinacol rearrangement.

propionic acid that was the starting material for this sequence, although for convenience one particular enantiomer is depicted in Figure 3.1. The amine **6** and ketone **7** prepared in this way were found to have the same sign of optical rotation, as was the case in the actual semipinacol rearrangement. Inspection of the structures for **6** and **7** thus reveals that the predominant stereochemistry in the rearrangement must have been inversion of configuration.

The quantitative results for the semipinacol rearrangement could be accounted for by a mixed mechanism consisting of 24.5% A and 75.5% B. However Collins recognized that there was a more compact way of describing the phenomenon. It is shown in Figure 3.2. The new mechanism combines the best of both worlds in that the condition $k_r \gg k_m$ would be indistinguishable from mechanism A while $k_m \gg k_r$ would be indistinguishable (at least by the criteria used here) from mechanism B. But the important point about the new hypothesis is that the condition $k_r \sim k_m$ results in a new situation that neither A nor B alone could account for, namely that the product would exhibit predominant inversion but would be $<100\%$ optically pure. This is, of course, the experimental observation. The quantitative results could be accounted for if $k_m/k_r = 6.3$.

In order to test the new hypothesis Collins and coworkers carried out a challenging combination of labeling and optical activity studies. They recognized that if one could prepare the optically active, labeled amine **8**, then the mechanism shown in Figure 3.2 should exhibit migration of the labeled phenyl group in the inversion product but migration of the unlabeled phenyl in the retention product. The ingenious synthesis of **8** is shown in Figure 3.3. By making a stannic chloride complex of the starting α-amino ketone, it was possible to use the methyl group to control the stereochemistry of the subsequent Grignard addition. The racemic **8** so produced was a single diastereomer. Resolution was achieved with camphorsulfonic acid, as before.

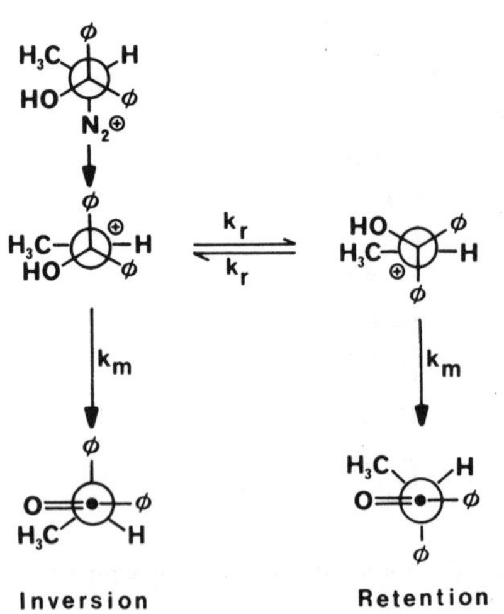

FIGURE 3.2. Semipinacol rearrangement by way of a carbonium ion that rotates and rearranges at comparable rates.

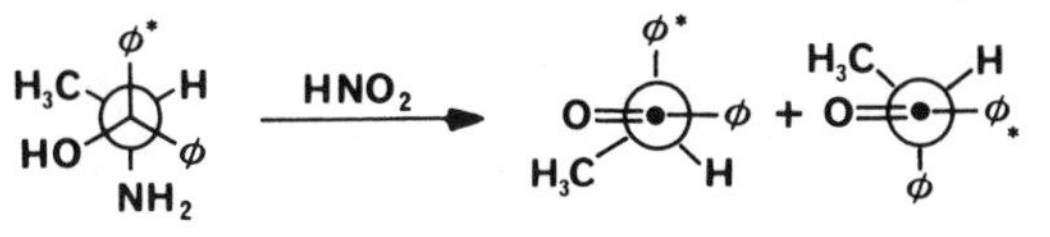

Analysis of the ^{14}C-label location in the product from the semipinacol rearrangement was carried out by nitrosation followed by base-promoted fragmentation.

Table 3.1 shows the experimentally determined specific radioactivities of the oxime and acid fragments (with respect to a specific activity of 100% for the reactant). The experiment was carried out on both the resolved sample of **8** and on a racemic sample. The predicted label distributions in Table 3.1 are based on the mechanism in Figure 3.2.

There is clearly good qualitative agreement between predicted and observed results in Table 3.1. The small quantitative discrepancy and the $>100\%$ total were eventually traced to a slow oxidation process that converted the aceto-phenone oxime to benzoic acid. The experimental data were thus completely consistent with the proposed mechanism.

If these experiments were carried out today one might use an optically active shift reagent to determine optical purities instead of the tedious procedure of recrystallizing to constant rotation. In addition, the isotopic label used in the final experiment might be ^{13}C rather than [14]C—its location could then be deter-

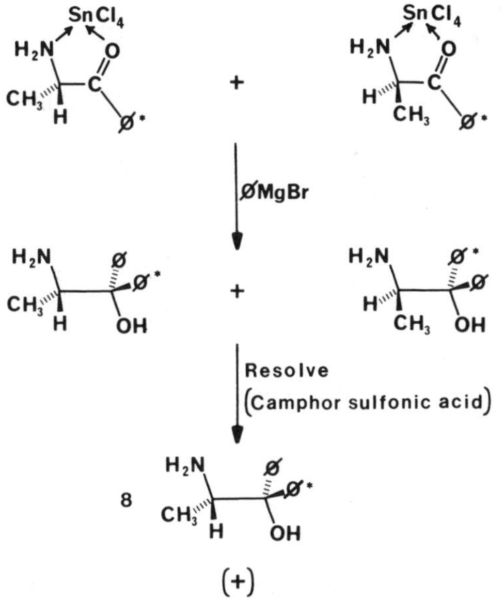

FIGURE 3.3. Synthesis of the optically active, labeled reactant for the semipinacol rearrangement. O* is a phenyl group labeled at the *ipso* carbon with [14]C.

mined by ^{13}C NMR. The fact that these techniques were not available to Collins and coworkers makes their achievements all the more impressive.

TABLE 3.1. Label Locations in the Products from Semipinacol Rearrangement of 8

		% Radioactivity	
		Oxime	Benzoic Acid
(+) 8	Predicted:	100%	0%
	Observed:	98%	4.4%
(±) 8	Predicted:	50%	50%
	Observed:	53%	44.3%

3.3. CHIRALITY CHANGES AT PRIMARY CARBON

A primary carbon is, by definition, an achiral center. The determination of intrinsic chirality changes at such a site thus requires that one of the methylene hydrogens be replaced by a heavier isotope—usually deuterium. Since only one of the prochiral methylene hydrogens can be labeled, this is often a formidable synthetic challenge. Clearly, no reagent would enable one to resolve a racemic mixture of label-isomeric enantiomers. Determination of optical purities is equally difficult. Rotations of molecules that rely on isotopic substitution for their optical activity are usually very small—often $< 1°$—and so contamination by optically active impurities can become a serious problem. In the following subsections we will see a number of ways of dealing with these difficulties.

3.3.1. Direct Measurement of Optical Rotations

Despite the problems, a number of studies involving direct measurement of optical rotations have been carried out on compounds that are chiral by virtue of an isotopic label. An interesting example comes from the work of Lown.[4] The 1,4-dihydropyridazine 9 rearranges to 10 upon warming to 55°C. A double-labeling crossover experiment showed that the reaction was 88% intramolecular and 12% intermolecular. The intermolecular component could be intercepted by adding *n*-butyl thiol, suggesting a radical mechanism. Lown chose to investigate the stereochemistry of the intramolecular process by stereospecific labeling of one of the benzylic hydrogens. The correlation of reactant and product configurations and the synthesis of the optically active starting material are shown in Figure 3.4.

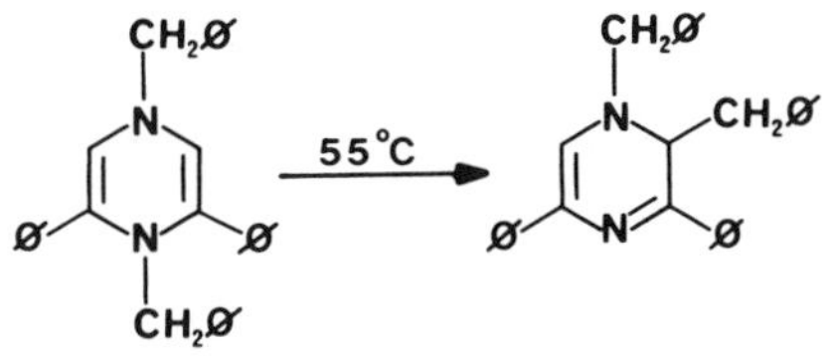

The synthetic task was made much easier by the fact that the starting material, (+)-benzylamine-α-*d*, was already known. In the correlation sequence the Stevens rearrangement was assumed to proceed with retention of configuration by analogy with the work of Schollkopf.[5] If the Stevens rearrangement was completely stereospecific the data indicate that the [1,3] benzyl migration must have occurred with nearly complete inversion—a result consistent with a Woodward–Hoffmann-allowed pericyclic mechanism.

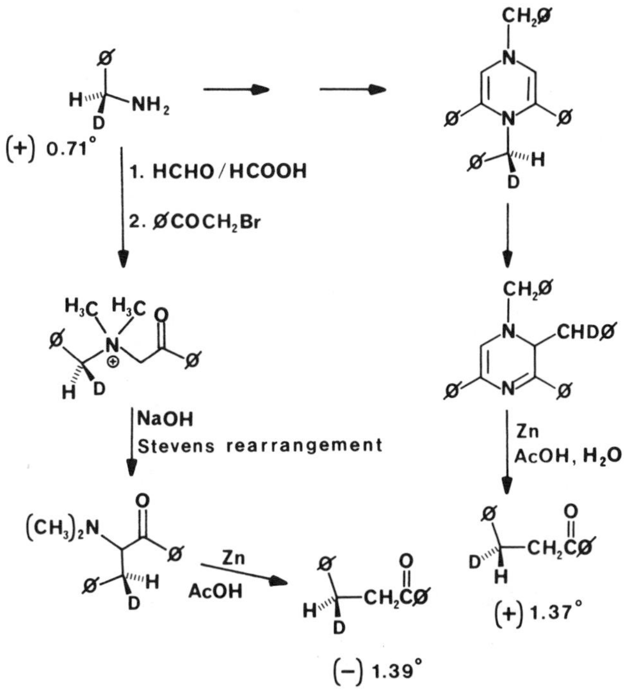

FIGURE 3.4. Correlation or reactant and product configurations for a [1,3] sigmatropic benzyl migration.

3.3.2. Use of an Existing Reference Chiral Center

One way to circumvent the resolution and analysis problems of reactions involving chiral primary centers is to make use of another chiral center within the molecule as a reference. This center must have a configuration that either remains fixed during the reaction or changes in a known manner. An example is the work of Zoeckler and Carpenter[6] on the alkoxide-accelerated [1,3] rearrangement. The potassium alkoxide **11** undergoes a rearrangement to **12** at

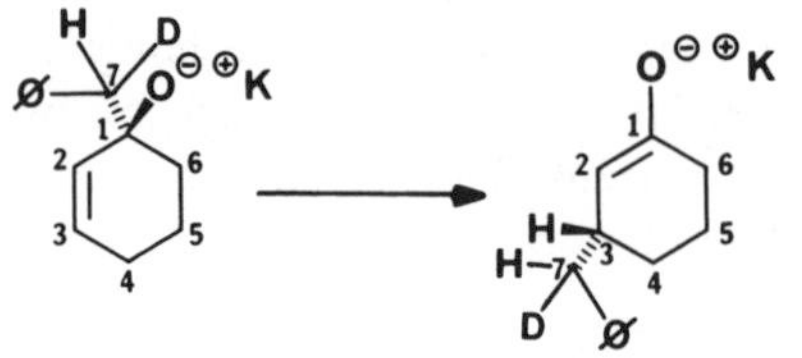

room temperature. A double-labeling crossover experiment showed that the reaction was 75% intramolecular. If one assumes that the intramolecular reaction would be constrained to be suprafacial then it becomes possible to determine the stereochemistry of the reaction at the migrating center without resorting to optical activity measurements. The relative chiralities of C1 and C7 in **11** could

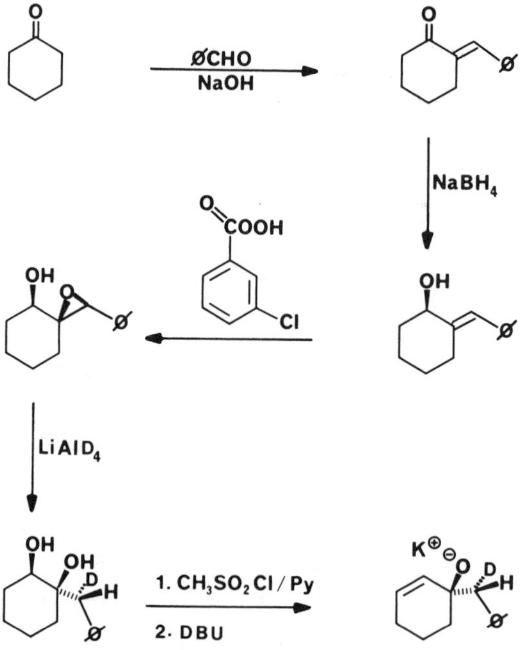

FIGURE 3.5. Preparation of the reactant for an alkoxide-accelerated]1,3] sigmatropic benzyl migration. All compounds in this figure are racemic but single diastereomers.

be fixed by synthesis; the relative chiralities of C1 in **11** and C3 in **12** are fixed by the suprafacial constraint, and so determination of the overall reaction stereochemistry reduces to a problem of determining the relative chiralities of C3 and C7 in **12**. The synthetic sequence that afforded a single diastereomer of racemic **11** is shown in Figure 3.5.

The final stereochemical analysis on **12** was achieved by conversion to a benzobicyclo[3.3.1]nonene in which the diastereotopic benzylic hydrogens exhibited a difference of almost 0.5 ppm in the proton NMR. Assignment of the resonances for these two hydrogens could be made by the difference in their coupling to the vicinal bridgehead hydrogen but it was confirmed by independent synthesis (see Figure 3.6.).

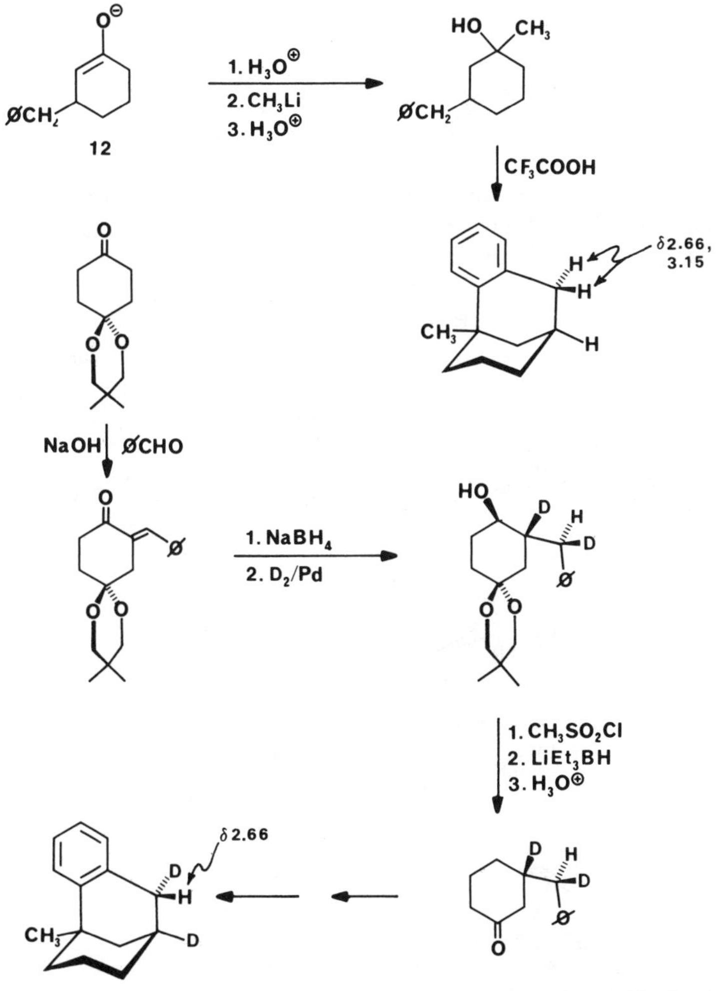

FIGURE 3.6. Assignment of the relative stereochemistries at the benzylic and vicinally related ring carbon in **12**. All compounds shown are racemic but single diastereomers.

The final result was that the [1,3] migration was found to proceed with at least 65% retention of configuration at the migrating carbon, in contrast to the result for the 1,4-dihydropyridazine rearrangement described in Section 3.3.1. A possible reason for the difference was advanced by the authors.

3.3.3. Creation of a Reference Chiral Center

When there is no convenient reference chiral center in a molecule it is sometimes possible to create a new one. The main concern about employing such a strategy is that the chirality of the new reference center might actually change the mechanism under investigation. One way to avoid such problems is to make sure that the reference chiral center, like the reaction center, is chiral by virtue of an isotopic difference. Thus for a reaction such as:

13 would not be a permissible model reactant since the new chiral center could easily influence the stereochemistry of the reaction at the adjacent site. One need have no such concerns about the isotopic analog **14**.

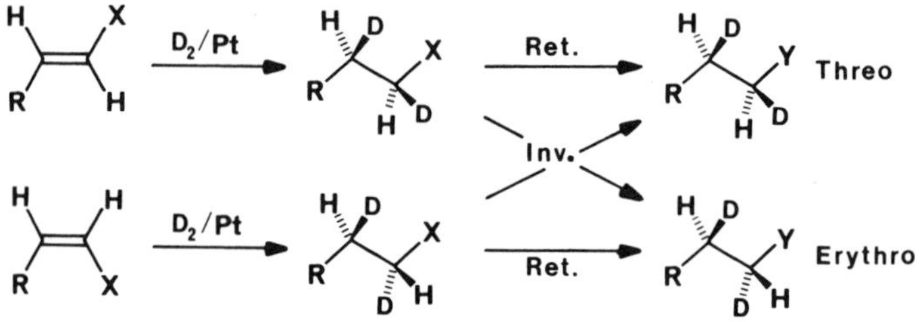

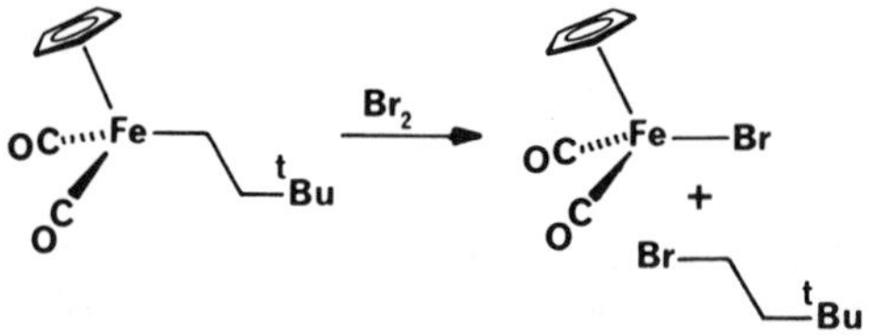

15

At first sight it might appear that the synthesis of **14**, with its two chiral isotopic centers, would be worse than the synthesis of an optically active compound labeled just at the reaction center. In fact this is not the case. Because it is only the *relative* stereochemistry of the two chiral centers in **14** that interests us, the compound can be prepared in racemic form. Both isotopic diastereomers are available through catalytic hydrogenation (or deuterogenation). The *erythro* and *threo* diastereomers can usually be identified by the difference in J_{HH} in the NMR spectrum.

The work of Whitesides and coworkers[7] on the electrophilic cleavage of iron–carbon σ-bonds is a nice illustration of the utility of this technique. Treatment

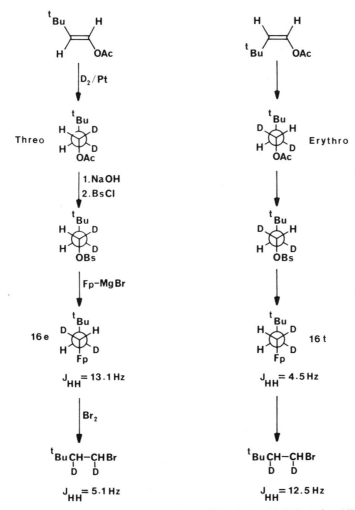

FIGURE 3.7. Determination of the stereochemistry of Fe—C cleavage by bromine. All compounds shown are racemic but single diastereomers. Fp is the group $(\eta^5\text{-}C_5H_5)Fe(CO)_2$. Bs is $p\text{-}BrC_6H_4SO_2\text{-}$.

of complexes such as **15** with bromine results in cleavage of the metal–carbon bond. In order to determine the stereochemistry at carbon of this process the isotopic diastereomers **16t** and **16e** were prepared as shown in Figure 3.7. **16t** Showed $J_{HH} = 4.5$ Hz in the NMR spectrum whereas **16e** had $J_{HH} = 13.1$ Hz. A larger coupling constant for the *erythro* diastereomer is expected on the basis of the Karplus equation and the dihedral angle between the vicinal hydrogens in the lowest energy rotamer.

The alkyl bromide produced by bromination of **16t** had $J_{HH} = 12.5$ Hz, while the bromide from **16e** had $J_{HH} = 5.1$ Hz. Thus it appears that the bond cleavage occurred with complete inversion of configuration at carbon.

3.3.4. The Chiral Methyl Group

The ultimate challenge in the realm of isotope-induced chirality is to create a single enantiomer of a molecule in which three different isotopes of the same element occupy homotopic sites. Remarkably, this feat has been accomplished for the methyl group with ^1H, ^2H, and ^3H and for the phosphate group with ^{16}O, ^{17}O, and ^{18}O. We will concentrate on the former.

Two syntheses of optically active acetic acid-α-d-α-t were reported in 1969[8,9] but perhaps the most efficient route so far is the one worked out by Arigoni and coworkers in 1975.[10] The synthesis is outlined in Figure 3.8.

This route provides access to both enantiomers of the doubly labeled acetic acid, which can then be converted to other methyl-containing compounds by the usual functional group manipulations.

Most applications of this chemistry have been in the bioorganic field, perhaps because the assay of the reaction products requires the use of enzymes. It seems probable, however, that applications to more traditional physical organic problems will be seen soon.

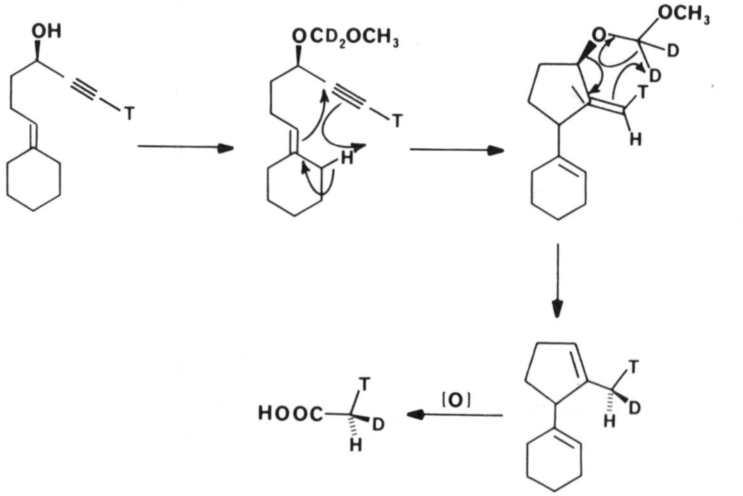

FIGURE 3.8. Route to optically active acetic acid-d,t worked out by Arigoni and coworkers.[10]

3.4. RACEMIZATION/EXCHANGE STUDIES

The combined application of isotope incorporation, chirality, and kinetic studies can result in a very powerful tool for studying reaction mechanisms. In this section we will consider two examples that concern the geometries of carbanions.

One normally expects carbanions that are stabilized by functional groups such as —COR, —CO$_2$R, or —CN to be sp^2 hybridized (i.e., trigonal planar) in order to maximize the overlap of p_π orbitals. When such a carbanion is incorporated into a cyclopropane ring, however, the situation is not so clear because the improved overlap that occurs on rehybridization from sp^3 to sp^2 is accompanied by an increase in ring strain (the ideal CCC angle being 109° for sp^3 but 120° for sp^2). There is no obvious way to predict which of these factors would be more important.

At first sight it might appear that the problem could be solved by treating the optically active cyclopropane **17** with base. If the resulting carbanion were planar the cyclopropane would racemize; if the anion were pyramidal the enantiomeric integrity would be retained. A little further thought shows two

serious flaws in this experiment, however. If the anion were pyramidal but capable of inversion, racemization could still occur. Worse still, if no racemization occurred it could simply be because the anion was never formed!

What is required is to determine how the rate of racemization compares with the rate of proton abstraction. The latter can be determined by treating the cyclopropane with base in a deuterated hydroxylic solvent. If the rate constant for racemization is k_α and that for isotope exchange is k_{ex}, then the predicted results for the three situations under consideration are:

Planar anion: $k_\alpha / k_{ex} = 1$

Pyramidal anion: $k_\alpha / k_{ex} < 1$

No reaction: $k_\alpha = k_{ex} = 0$

The experiment was conducted by Walborsky and coworkers[11] in 1962. Their result of $k_\alpha/k_{ex} = 1.24 \times 10^{-4}$ (for R = Ph with NaOMe/MeOH at 50°C) showed clearly that the anion was pyramidal but capable of slow inversion.

A more challenging, and still unresolved, problem to which this technique has been applied is the question of the geometry of α-sulfonyl carbanions. It is known that the group —SO_2R does stabilize carbanions but the mechanism by which it does so has been a matter for some debate. If one wished to invoke p_π–d_π overlap, utilizing the empty sulfur d orbitals, then the carbanion would presumably have to be sp^2 hybridized. On the other hand, a simple inductive stabilization would presumably allow the anion to adopt the usual pyramidal geometry. Corey and coworkers[12] found that the sulfone **18** exhibited $k_\alpha/k_{ex} = 0.024$, which, by comparison with the previous example, might lead one to believe that the anion must be pyramidal. In fact this need not be so. A planar anion resulting from deprotonation of **18** would still be chiral unless it were able

Hx = n-hexyl

to rotate about the S—C⁻ bond. But restricted rotation about this bond might be just what one would expect for an anion experiencing p_π–d_π bonding!

Further light was cast on the problem by a later experiment of Corey and Lowry[13] (see Figure 3.9). The cyclic sulfone **19** was prepared from optically active **20** whose absolute configuration was known. Upon treatment with base a fragmentation occurred to give sulfinate anion **21**, which was alkylated with benzyl bromide. The resulting sulfone (**22**) was found to be optically pure and levorotatory when the R(+) enantiomer of **19** was used. When **22** was independently synthesized from S(−)-1-chloro-1-phenylethane it was found to be dextrorotatory, indicating that the fragmentation of **19** had occurred with complete inversion of configuration. This result shows that protonation of the α-sulfonyl carbanion intermediate must have occurred faster than rotation about the S—C⁻ bond but slower than inversion at the carbanionic center. Such a result would be consistent with a planar anion but could also be explained as a pyramidal anion that could be protonated only from the side of the sulfonyl oxygens:

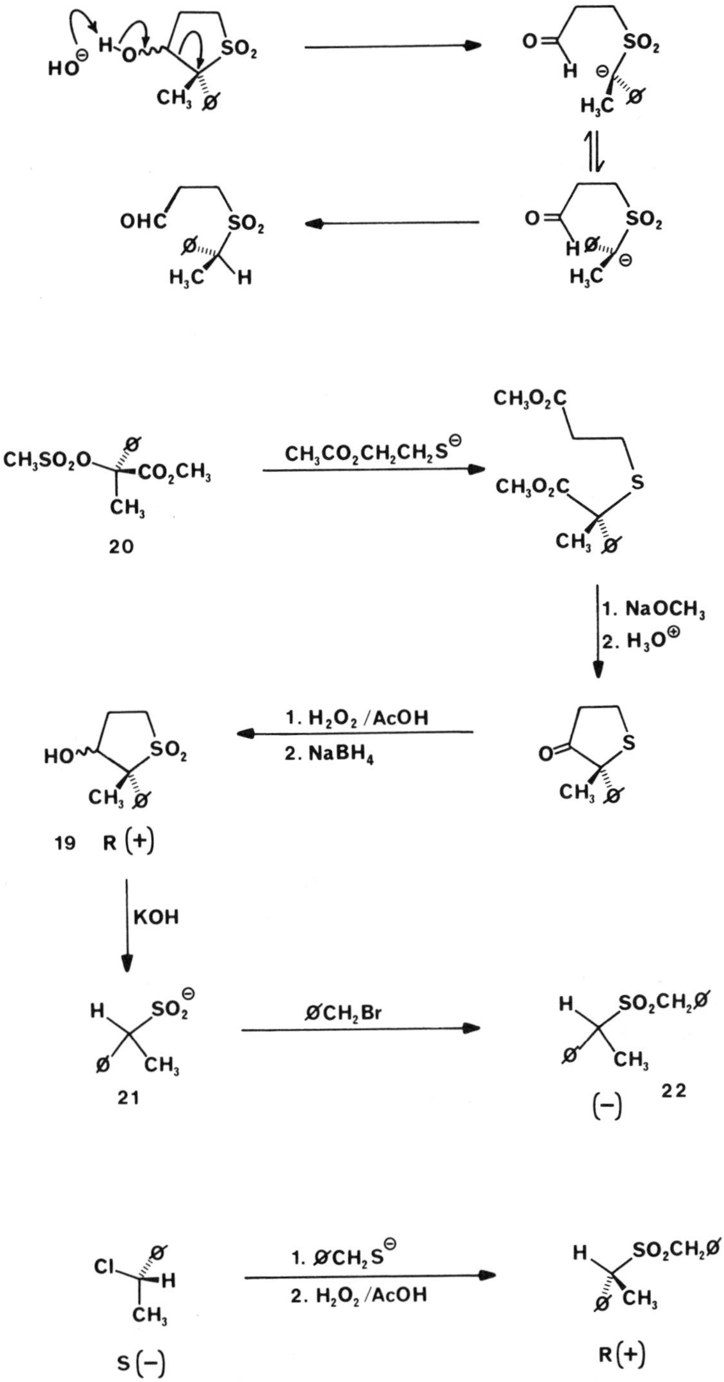

FIGURE 3.9. Correlation of reactant (19) and product (22) configurations for the generation and protonation of an α-sulfonyl anion.

The true structure of α-sulfonyl carbanions remains unknown. *Ab initio* molecular orbital calculations on the model $HSO_2CH_2^-$ show a lower energy for the pyramidal structure.[14]

3.5. TOLBERT ANALYSIS

Perhaps one of the most intractable problems in modern physical organic chemistry has been the differentiation between pericyclic and biradical mechanisms for the thermal transformations such as [1,3] and [3,3] sigmatropic shifts, cyclobutane fragmentation, and the Diels–Alder reaction. We will see many attempts to resolve this problem in later chapters. This section deals with the information that can be gained by a special type of chirality study.

The Diels–Alder reaction of diphenylisobenzofuran with dimethyl fumarate could be imagined to proceed by a pericyclic ($_\pi2_s + _\pi4_s$) mechanism (A) or to involve a biradical intermediate (B).

Tolbert and Ali have attempted to distinguish between these two mechanisms by use of optically active dienophiles.[15] The three dienophiles **23a–c** were each reacted with diphenylisobenzofuran to give the products shown in Figure 3.10. With the aid of a few approximations one can make predictions about the ratios of these products for each of the mechanisms.

For mechanism A one can write:

$$\Delta G_\alpha^\ddagger = \Delta G_o^\ddagger + (\Delta G_{\alpha\text{-}endo}^\ddagger - \Delta G_o^\ddagger) + (\Delta G_{\alpha\text{-}exo}^\ddagger - \Delta G_o^\ddagger)$$

$$= \Delta G_{\alpha\text{-}endo}^\ddagger + \Delta G_{\alpha\text{-}exo}^\ddagger - \Delta G_o^\ddagger$$

where $\Delta G_i^\ddagger$ is the free energy of activation for formation of product i. This equation is based on the assumption of additivity of substituent effects, that is, replacement of methyl by *l*-bornyl is assumed to have a constant effect on $\Delta G^\ddagger$ regardless of the chemical constitution of the other ester group. Different effects

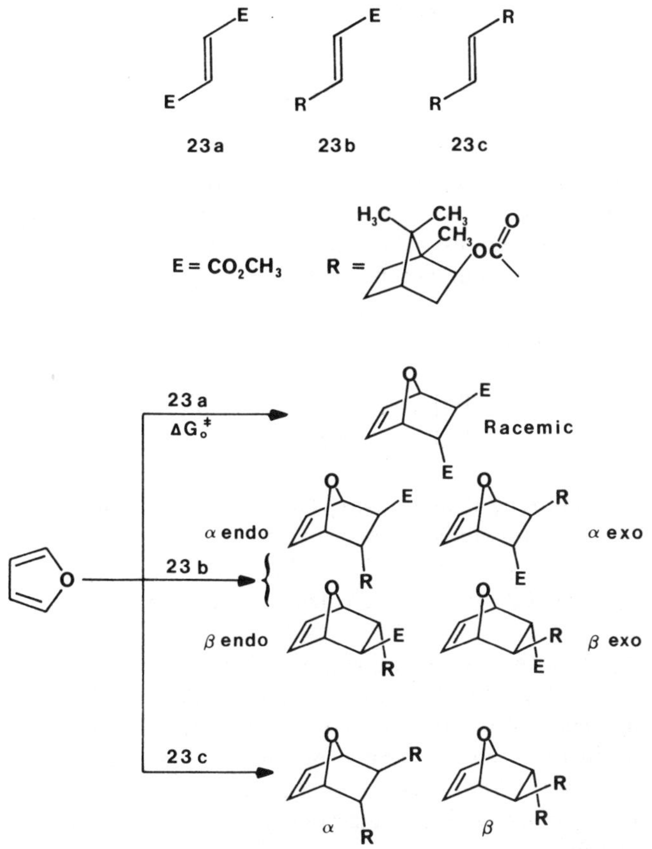

FIGURE 3.10. Products from the reaction of the dienophiles **23a-c** with diphenylisobenzofuran. Only the furan part of the diene is shown.

on $\Delta G^{\ddagger}$ for replacements at different sites (i.e., *endo* or *exo*, α or β) are allowed, however.

By analogy,

$$\Delta G_{\beta}^{\ddagger} = \Delta G_{\beta\text{-}endo}^{\ddagger} + \Delta G_{\beta\text{-}exo}^{\ddagger} - \Delta G_{o}^{\ddagger}$$

Hence,

$$\Delta G_{\alpha}^{\ddagger} - \Delta G_{\beta}^{\ddagger} = \Delta G_{\alpha\text{-}endo}^{\ddagger} + \Delta G_{\alpha\text{-}exo}^{\ddagger} - \Delta G_{\beta\text{-}endo}^{\ddagger} - \Delta G_{\beta\text{-}exo}^{\ddagger}$$

From the Eyring equation,

$$\Delta G_{i}^{\ddagger} = RT \ln(\mathbf{k}T/h) - RT \ln k_{i}$$

where $\mathbf{k}$ = Boltzmann's constant
        h = Planck's constant
        k = rate constant

Hence,

$$\frac{k_\alpha}{k_\beta} = \frac{k_{\alpha\text{-endo}} k_{\alpha\text{-exo}}}{k_{\beta\text{-endo}} k_{\beta\text{-exo}}} \tag{1}$$

where k is the rate constant for formation of product i.

For mechanism B the substituent effects should not be additive since the new C—C bonds are formed nonsynchronously. Hence Eq. (1) might be expected to hold for mechanism A but not for mechanism B. The experimental effort required to perform this test is relatively modest since the rate constant ratios of Eq. (1) can be replaced by concentration ratios, giving Eq. (2).

$$\frac{[\alpha]}{[\beta]} = \frac{[\alpha\text{-}endo]}{[\beta\text{-}endo]} \cdot \frac{[\alpha\text{-}exo]}{[\beta\text{-}exo]} \tag{2}$$

Tolbert and Ali found that $[\alpha\text{-}endo]/[\beta\text{-}endo] = 1.53$ and that $[\alpha\text{-}exo]/[\beta\text{-}exo] = 1.41$, leading to a predicted value of 2.16 for $[\alpha]/[\beta]$ if mechanism A were correct. The experimental value for $[\alpha]/[\beta]$ was 2.08, which the authors took to be within experimental error of the calculated value, thereby supporting the pericyclic mechanism.

Two problems with this technique are evident. The first is a difficulty that we saw in chapter 1, namely that a mathematical equality [Eq. (2) in this case] is never truly reflected in an experimental determination because of experimental error. In the present example one has to presume that the observed $[\alpha]/[\beta]$ of 2.08 was *within experimental error* of the calculated value, 2.16. Even the latter number has some uncertainty since it was derived from experimental measurements.

The second problem is intimately related. At present there is no way to judge how large a deviation from Eq. (2) would be expected for mechanism B. One cannot be certain that the difference between 2.08 and 2.16 is too small to exclude mechanism B.

The problems might be resolved by applying the Tolbert analysis to a variety of reactions and looking for a dichotomy of results—one class of reactions in which the observed product ratio is within experimental error of the calculated ratio, and a second class in which there is (statistically) significant deviation. A successful calibration of this kind could lead to the development of a very powerful tool for mechanistic study.

We will see a closely related technique that employs isotope effects when we discuss the Thornton analysis in Chapter 5.

3.6. THE SKELL HYPOTHESIS

In 1956 Skell and coworkers postulated that the cycloaddition of carbenes to olefins should be stereospecific if the carbene were in a singlet state but nonstereospecific if it were in a triplet state. Their reasoning was that for the singlet carbene, concerted formation of the two new C—C bonds would be allowed whereas for the triplet carbene a biradical intermediate, which could lose stereochemistry by bond rotation, would necessarily be involved.[16]

If the hypothesis were correct one could use the stereospecificity of cyclo-addition to determine the spin states of other reactive intermediates. In fact this kind of reasoning has been applied to the thermal reactions of nitrenes,[17] trimethylenemethane biradicals,[18] and cyclobutadiene.[19] It is also common in the discussion of photochemical cycloadditions such as the Paterno–Buchi reaction.[20]

A serious blow to the Skell hypothesis was dealt by Schuster and coworkers[21] in 1981 when they reported that singlet fluorenylidene **24**, generated by laser flash photolysis of 9-diazofluorene, exhibited *nonstereospecific* cycloaddition to *cis*-β-methylstyrene and *cis*-2-pentene.

Subsequently Platz and coworkers[22] have shown that the reactive intermediate in Schuster's experiment was probably the triplet state of fluorenylidene and not the singlet as Schuster had thought. Nevertheless, the incident serves as a useful reminder that there is no rigorous requirement for adherence to the Skell hypothesis and that conclusions about the spin state of reactive intermediates based on cycloaddition stereochemistry are consequently subject to some uncertainty.

REFERENCES

1. A. G. Schultz, J. J. Napier, and R. Lee, *J. Org. Chem.*, **44**, 663 (1979)

2. F. R. Jensen, *J. Am. Chem. Soc.*, **82**, 2469 (1960).

3. B. M. Benjamin, H. J. Schaeffer, and C. J. Collins, *J. Am. Chem. Soc.*, **79**, 6160 (1957).

4. J. W. Lown, M. Akhtar, and R. S. McDaniel, *J. Org. Chem.*, **39**, 1998 (1974).

5. J. H. Brewster and M. W. Kline, *J. Am. Chem. Soc.*, **74**, 5179 (1952); U. Schollkopf, U. Ludwig, G. Osterman, and M. Patsch, *Tetrahedron Lett.*, 3415 (1969).

6. M. T. Zoeckler and B. K. Carpenter, *J. Am. Chem. Soc.*, **103**, 7661 (1981).

7. P. L. Bock, D. J. Boschetto, J. R. Rasmussen, J. P. Demers, and G. M. Whitesides, *J. Am. Chem. Soc.*, **96**, 2814 (1974).

8. J. W. Cornforth, J. W. Redmond, H. Eggerer, W. Buckel, and C. Gutschow, *Nature*, **221**, 1212 (1969).

9. J. Lüthy, J. Rétey, and D. Arigoni, *Nature*, **221**, 1213 (1969).

10. C. A. Townsend, T. Scholl, and D. Arigoni, *J. Chem. Soc. Chem. Commun.*, 921 (1975).

11. H. M. Walborsky, A. A. Youssef, and J. M. Motes, *J. Am. Chem. Soc.*, **84**, 2465 (1962).

12. E. J. Corey and E. T. Kaiser, *J. Am. Chem. Soc.*, **83**, 490 (1961).

13. E. J. Corey and T. H. Lowry, *Tetrahedron Lett.*, 793 (1965).

14. S. Wolfe, A. Rauk, and I. G. Csizmadia, *J. Am. Chem. Soc.*, **91**, 1567 (1969).

15. L. M. Tolbert and B. Ali, *J. Am. Chem. Soc.*, **103**, 2104 (1981).

16. P. S. Skell and R. C. Woodworth, *J. Am. Chem. Soc.*, **78**, 4496 (1956).

17. W. Lwowski and F. P. Woerner, *J. Am. Chem. Soc.*, **87**, 5491 (1965); J. S. McConaghy and W. Lwowski, *Ibid.*, **89**, 2357 (1967).

18. L. R. Corwin, D. M. McDaniel, R. J. Bushby, and J. A. Berson, *J. Am. Chem. Soc.*, **102**, 276 (1980).

19. L. Watts, J. D. Fitzpatrick, and R. Pettit, *J. Am. Chem. Soc.*, **88**, 623 (1966).

20. D. R. Arnold, R. L. Hinman, and A. H. Glick, *Tetrahedron Lett.*, 1425 (1964); N. C. Yang and W. Eisenhardt, *J. Am. Chem. Soc.*, **93**, 1277 (1971).

21. J. J. Zupancic and G. B. Schuster, *J. Am. Chem. Soc.*, **103**, 944 (1981).

22. D. Griller, C. R. Montgomery, J. C. Scaiano, M. S. Platz, and L. Hadel, *J. Am. Chem. Soc.*, **104**, 6813 (1982).

CHAPTER 4

KINETICS

Kinetic studies, in one guise or another, have historically made the greatest contribution to the elucidation of reaction mechanisms. In addition to the determination of reaction orders and rate constants, which will be the topics of this chapter, racemization/exchange studies (Chapter 3), isotope effects (Chapter 5), and the interpretation of activation parameters (Chapter 7) all rely heavily on kinetic measurements.

The experimental methods for determining reaction kinetics and the mathematical methods for handling the data have been described in a number of excellent specialist texts[1] and so will be treated relatively briefly here. The application of these classical kinetic methods will be illustrated by a fairly detailed analysis of the development of ion pairing concepts in solvolysis reactions. In addition a powerful mathematical technique for handling arrays of reversible unimolecular reactions will be developed and illustrated. Finally the topic of numerical integration of kinetic rate laws will be discussed.

4.1. INTEGRATED FORMS OF SIMPLE RATE EXPRESSIONS

First-order, irreversible reaction:

$$A \xrightarrow{k} B$$

$$\frac{-d[A]}{dt} = k[A]$$

$$\int [A]^{-1} d[A] = -k \int dt$$

$$\ln [A] - \ln [A]_0 = -kt$$

or

$$[A] = [A]_0 e^{-kt}$$

The rate constant k is determined by plotting a graph of $\ln [A]$ vs. t.
First-order, reversible reaction

$$A \underset{k_r}{\overset{k_f}{\rightleftharpoons}} B$$

Assuming the $[A] = [A]_0$, $[B] = 0$ at $t = 0$:

$$[A] + [B] = [A]_0 \text{ at all } t$$

$$\frac{-d[A]}{dt} = k_f[A] - k_r[B]$$

$$= (k_f + k_r)[A] - k_r[A]_0$$

Integrating

$$\ln \left\{ \frac{k_f[A]_0}{(k_f + k_r)[A] - k_r[A]_0} \right\} = (k_f + k_r)t$$

At equilibrium

$$k_f[A]_e = k_r\{[A]_0 - [A]_e\}$$

$$[A]_e = \frac{k_r[A]_0}{k_f + k_r} \tag{1}$$

$$\ln \left\{ \frac{[A]_0 - [A]_e}{[A] - [A]_e} \right\} = (k_f + k_r)t \tag{2}$$

$(k_f + k_r)$ can be determined from Eq. (2) and then individual values of k_f and k_r found from Eq. (1).
Second-order, irreversible reaction (dimerization, etc.):

$$2A \overset{k}{\longrightarrow} B$$

$$\frac{-d[A]}{dt} = 2k[A]^2$$

The factor of 2 in the rate law is often omitted (i.e., incorporated into k) but this can lead to confusion when the dimerization is part of a larger scheme of reactions. It is retained explicitly in the following. The integration is easy and gives

$$[A]^{-1} = [A]_0^{-1} + 2kt$$

Second-order, irreversible reaction (first order in each of A and B):

$$A + B \xrightarrow{k} C$$

$$\frac{-d[A]}{dt} = \frac{-d[B]}{dt}$$

$$[A] - [A]_0 = [B] - [B]_0$$

$$\frac{-d[A]}{dt} = k[A][B]$$

$$= k[A]\{[A] - [A]_0 + [B]_0\}$$

Integrating

$$\frac{1}{[B]_0 - [A]_0} \ln\left\{\frac{[A]_0([B]_0 - [A]_0 + [A])}{[A][B]_0}\right\} = kt$$

Note that if $[B]_0 \gg [A]_0$ (and hence $[B]_0 \gg [A]$):

$$\frac{1}{[B]_0} \ln\left\{\frac{[A]_0}{[A]}\right\} = kt$$

This is identical with the expression for a first-order reaction except for the factor of $[B]_0^{-1}$. The condition $[B]_0 \gg [A]_0$ leads to so-called pseudo-first-order kinetics and provides a convenient way to determine k.

4.2.   THE STEADY-STATE APPROXIMATION

In a reaction of the type

$$A \underset{k_{BA}}{\overset{k_{AB}}{\rightleftharpoons}} B \xrightarrow{k_{BC}} C \tag{3}$$

if $k_{BA} \gg k_{AB}$ or $k_{BC} \gg k_{AB}$, the kinetic analysis can be simplified considerably by setting $d[B]/dt = 0$. The rationale is that, under the circumstances specified, the change in concentration of B is negligible compared to the changes in concentration of A and C. (The *percentage* change in [B] might be substantial but the *absolute* change of concentration per unit time is small.) After making the steady-state approximation it is easily shown that the rate law becomes:

$$\frac{-d[A]}{dt} = \frac{k_{AB}k_{BC}[A]}{k_{BA} + k_{BC}} \tag{4}$$

The accuracy of this approximation can be illustrated by an example. Suppose $[A]_0 = 1$; $[B]_0 = [C]_0 = 0$ and $k_{AB} = 10^{-5}$ sec^{-1}; $k_{BA} = 2 \times 10^{-5}$ sec^{-1}; $k_{BC} = 2 \times 10^{-4}$ sec^{-1}. The time evolution of [A] can now be calculated by using the steady-state approximation and the result compared with the exact solution (for which see reference 1). The results are presented in Table 4.1.

The steady-state approximation can be applied to reactions of any molecularity. When the intermediate reacts by nonunimolecular processes the rate law will have concentration terms in the denominator of a fraction. For example,

$$A \underset{k_{BA}}{\overset{k_{AB}}{\rightleftharpoons}} B \overset{k_{BD}[C]}{\longrightarrow} D$$

has the rate law

$$\frac{-d[A]}{dt} = \frac{k_{AB}k_{BD}[A][C]}{k_{BA} + k_{BD}[C]}$$

TABLE 4.1. Comparison of Exact Solution and Steady-State Approximation for Time Evolution of [A] in Eq. (3)

t(sec)	[A](SSA)[a]	[A](Exact)	% Error
100	0.9991	0.9990	0.01
500	0.9955	0.9950	0.05
1,000	0.9910	0.9901	0.09
5,000	0.9556	0.9530	0.27
10,000	0.9131	0.9099	0.35
50,000	0.6347	0.6331	0.25
100,000	0.4029	0.4027	0.05
500,000	0.0106	0.0108	1.85

[a]Concentration of A calculated by applying the steady-state approximation.

4.3. MECHANISTIC INTERPRETATION OF RATE LAWS

A particular mechanism for a reaction always defines a unique rate law for the kinetics (even though it might be difficult to determine what the equation is in certain cases). Regrettably, the converse is not true. Nevertheless, it is possible to determine something about the general features of a mechanism from an observed rate law.

(a)   For a single step reaction a rate law of the form

$$\text{Rate} = k[\text{A}]^a[\text{B}]^b \cdots$$

defines a transition state of chemical composition $a\,\text{A} + b\,\text{B} + \cdots$.

This conclusion can be derived from the transition state theory model for chemical reactions.[1]

$$a\,\text{A} + b\,\text{B} + \cdots \rightleftharpoons \text{TS} \tag{5}$$

In Eq. (5) the reactants A, B, and so on, are assumed to be in equilibrium with the activated complex TS. The equilibrium constant, K, is defined in the usual way:

$$K = \frac{[\text{TS}]}{[\text{A}]^a[\text{B}]^b \cdots}$$

Hence

$$[\text{TS}] = K[\text{A}]^a[\text{B}]^b \cdots$$

According to transition state theory the rate of the reaction is proportional to the concentration of the activated complex:

$$\text{Rate} = c[\text{TS}]$$

where c is a constant of proportionality. Hence,

$$\text{Rate} = cK[\text{A}]^a[\text{B}]^b \cdots$$

which then has the form of the empirical rate law.

(b)   A rate law of the form

$$\text{Rate} = k[\text{A}][\text{B}] + k'[\text{A}][\text{B}][\text{C}] \tag{6}$$

indicates the existence of parallel pathways (i.e., competitive reactions) leading to transition states of composition A + B and A + B + C. Note that parallel pathways leading to transition states of the same composition do not result in a summation of terms since the rate constants for such pathways will simply be lumped together and appear as a single observed rate constant. Hence Eq. (6) would require a mechanism involving *at least* two different reactions for the disappearance of A and B.

(c)   A summation of terms in the denominator of a rate equation indicates the existence of at least one intermediate that is formed reversibly. The composition of the transition states leading to and from the intermediate(s) can be determined by considering each denominator term in turn to be dominant. Thus for a reaction A + 2B $\longrightarrow$ C, a rate law of the form:

$$\text{Rate} = \frac{k[A][B]^2}{1 + k'[A]}$$

indicates a reaction with at least two sequential steps, one having a transition state with chemical composition 2B, the other with a transition state of composition A + 2B. Given this information, one can see that the mechanism defined by Eqs. (7) and (8) would fit whereas that defined by Eqs. (9) and (10) would not. (This assertion can be checked by applying the steady-state approximation to the two schemes.)

$$B + B \rightleftharpoons B_2 \qquad\qquad (7)$$

$$A + B_2 \longrightarrow C \qquad\qquad (8)$$

$$A + B \rightleftharpoons AB \qquad\qquad (9)$$

$$AB + B \longrightarrow C \qquad\qquad (10)$$

4.4.   SOLVOLYSIS AND ION PAIRS

The application of the ion pairing concept to the solvolysis of alkyl halides by Winstein and coworkers represents an excellent illustration of the use of kinetic studies to elucidate reaction mechanisms. The evolution of increasingly detailed models in this work is also a nice demonstration of the "iterative" approach to problem solving in empirical science, as discussed in Chapter 1.

An extremeley important contribution to the understanding of solvolysis reactions was made by Ingold[2] in 1933 when he recognized the possibility of two classes of mechanism for nucleophilic substitution reactions. He gave these classes the now-familiar designations S_N1 and S_N2, where the S stands for "substitution", the N for "nucleophilic" and the number indicates the molecularity of the rate-determining step.

$$S_N1: \quad R\!-\!X \xrightarrow[\text{slow}]{} R^+ + X^- \xrightarrow[\text{fast}]{\text{Nu:}} R\!-\!Nu^+ + X^-$$

$$\frac{-d[RX]}{dt} = k[RX]$$

$$S_N2: \quad R\!-\!X + Nu: \longrightarrow R\!-\!Nu^+ + X^-$$

$$\frac{-d[RX]}{dt} = k[RX][Nu]$$

The difference in rate law for these two classes of substitution mechanism provides an experimental way of distinguishing between them in most cases. Unfortunately, solvolysis reactions are an exception. In the case where Nu: is the solvent the rate equation for S_N2 mechanism becomes:

$$\frac{-d[RX]}{dt} = k[RX][\text{solvent}]$$

$$= k'[RX]$$

which is experimentally indistinguishable from the rate equation for the S_N1 process.

A further complication presented by solvolysis reactions is that the solvent is usually a rather weak nucleophile and so the counterion, X^-, can be expected to compete effectively for the carbonium ion in an S_N1 reaction. In other words one has to allow for the possibility of reversibility in the first step:

$$R\!-\!X \underset{k_{-1}}{\overset{k_1}{\rightleftharpoons}} R^+ + X^- \xrightarrow[R'OH]{k_2} R\!-\!OR' + H^+ + X^-$$

This simple change has a substantial effect on the rate equation and hence on the expected experimental observations for the reaction:

$$\frac{-d[RX]}{dt} = k_1[RX] - k_{-1}[R^+][X^-]$$

$$\frac{d[R^+]}{dt} = k_1[RX] - k_{-1}[R^+][X^-] - k_2[R^+]$$

$$= 0 \text{ (steady-state approximation)}$$

$$[R^+] = \frac{k_1[RX]}{k_2 + k_{-1}[X^-]}$$

$$\frac{-d[RX]}{dt} = \frac{k_1 k_2}{k_2 + k_{-1}[X^-]} \cdot [RX] \qquad (11)$$

k_2 is a pseudo-first-order rate constant that contains the concentration of solvent:

$$k_2 = k_2'[R'OH]$$

Eq. (11) allows us to make the following predictions:

1. Since X^- is produced in the reaction, a first-order plot of $\ln\{[RX]_0/[RX]\}$ vs. t should show downward curvature rather than being linear.
2. Added X^- should decrease the observed rate constant. This phenomenon is called common ion rate depression.

Hughes and Ingold[3] showed that both predictions were confirmed for the solvolysis of 4,4'-dimethylbenzhydryl chloride (**1**) (see Figure 4.1). Note that the line corresponding to **1** + LiCl shows little curvature because the Cl$^-$ produced during the reaction now makes an insignificant contribution to the total chloride ion concentration.

The solvolysis of benzhydryl chloride (**2**) does not seem to be so well behaved. As shown in Figure 4.1, the first-order plot for disappearance of **2** shows no obvious curvature and addition of LiCl actually causes a slight increase in rate. Worse still is the solvolysis of t-butyl bromide (**3**), which shows *upward* curvature for the first-order plot and a substantial increase in rate upon addition of LiBr.

A clue to the cause of this behavior is provided by the graph showing the effect of LiBr on the solvolysis of **1**. The added salt causes an increase in reaction rate and a more pronounced downward curvature of the first-order plot. The explanation is that there is a salt effect[4] in which the increased ionic strength of the medium increases the propensity of RX to dissociate, by decreasing the activity coefficients of the resulting ions. Thus addition of a *common ion* salt to a solvolysis reaction has two opposing effects: common ion rate depression and salt-effect rate enhancement. For **1** the former is dominant, for **2** they are closely balanced, and for **3** the latter is dominant. In order to understand better why the different substrates show different responses to added common ion salts we will restate the above conclusion in mathematical terms.

Empirically one finds that normal salt effects appear as a linear dependence of the rate constant for dissociation on the concentration of ions:

$$k_1 = k_1^\circ \{1 + b[X^-]\} \qquad (12)$$

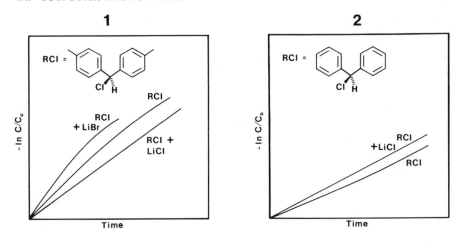

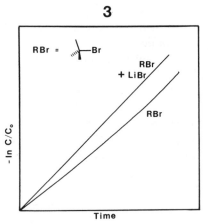

FIGURE 4.1. Solvolyses of 4,4′-dimethylbenzyhydryl chloride (**1**), benzhydryl chloride (**2**), and *tert*-butyl bromide (**3**). The data are plotted in the form that would be appropriate for a first-order irreversible reaction.

where b is some constant. Incorporating this into the previous rate expression [Eq. (11)] gives

$$\frac{-d[\text{RX}]}{dt} = \frac{k_1^{\circ} k_2 (1 + b[\text{X}^-])}{k_{-1}[\text{X}^-] + k_2} \cdot [\text{RX}]$$

If $k_{-1}[\text{X}^-] \gg k_2$,

$$\frac{-d[\text{RX}]}{dt} = \frac{k_1^{\circ} k_2}{k_{-1}} \left\{ \frac{1}{[\text{X}^-]} + b \right\} \cdot [\text{RX}]$$

which will lead to common ion rate depression and downward curvature of the first-order plot.

If $k_2 \gg k_{-1}[X^-]$,

$$\frac{-d[RX]}{dt} = k_1^{\circ}\,\{1 + b[X^-]\} \cdot [RX]$$

which will lead to upward curvature of the first-order plot and salt-effect rate enhancement.

If $k_{-1}[X^-] \gg k_2$ then $k_{-1}[X^-] \gg k_2'[R'OH]$, hence $k_{-1} \gg k_2'$ since $[R'OH] \gg [X^-]$. On the other hand, if $k^{-1}[X^-] \ll k_2$ then $k_{-1} \simeq k_2'$. (It is unlikely that $k_{-1} \ll k_2'$ since X^- is charged whereas $R'OH$ is not.) Hence our effort to understand the experimental results of Hughes and Ingold now reduces to a problem of determining why $k_{-1} \simeq k_2'$ for some carbonium ions but $k_{-1} \gg k_2'$ for others.

A plausible answer is illustrated in Figure 4.2. The Hammond postulate[5] says that the transition state for a reaction will most resemble in structure the intermediate to which it is closest in energy. In the present case, the transition state for addition of Nu to R^+ will look very much like R^+ if the reaction is highly exothermic but will take on more $R-Nu^+$ character as the reaction becomes less exothermic. Two extreme examples are illustrated in Figure 4.2. The sensitivity of the transition state to nucleophile structure (and hence the ability of R^+ to discriminate between two different nucleophiles) should therefore increase as the reactivity of the carbonium ion decreases. The expected order of reactivities for the carbonium ions studied by Hughes and Ingold is t-butyl > benzhydryl > 4,4'-dimethylbenzhydryl, providing a satisfyingly consistent explanation for the observed effects of common ion salt addition.

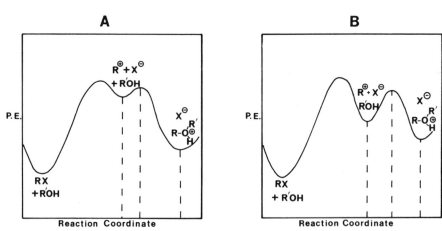

FIGURE 4.2. Potential energy profiles for formation and reaction of a highly reactive carbonium ion (A) and a relatively unreactive carbonium ion (B). Note the later (more productlike) transition state in the latter case.

It is perhaps worth summarizing the logical steps involved in this study since similar reasoning will be employed in the discussion that follows.

(a)   Additions of nucleophiles to reactive carbonium ions are, by necessity, highly exothermic and hence, by the Hammond postulate, should have transition states that depend little on nucleophile structure. As a consequence, the ability of reactive carbonium ions to distinguish between different nucleophiles is poor. In a solvolysis reaction this means that the rate constants for addition of X^- and $R'OH$ will be similar in magnitude, that is, $k_{-1} \simeq k_2'$, hence $k_{-1} [X^-] \ll k_2$. This leads to upward curvature of the first-order plot and salt-effect rate enhancement.

(b)   Nucleophilic additions to unreactive carbonium ions should have transition states that are more productlike and hence more sensitive to nucleophile structure. In a solvolyis reaction this means that the anionic nucleophile X^- should have a much higher intrinsic reactivity than the neutral nucleophile $R'OH$, that is, $k_{-1} \gg k_2'$, hence $k_{-1} [X^-] \gg k_2$ (at least in extreme cases). This leads to downward curvature of the first-order plot and to common ion rate depression.

The Hughes and Ingold model provided an internally consistent and apparently complete description of solvolysis reactions until the experiments of Winstein and coworkers that began in the 1950s.

In 1951 Winstein reported the solvolyses of 1-chloro-3-methyl-2-butene (4) and 3-chloro-3-methyl-1-butene (5) in acetic acid buffered with sodium acetate.[6] The sodium acetate served to keep the ionic strength of the medium constant throughout the reaction, as well as suppressing halide ion return (AcO⁻ being a more powerful nucleophile than Cl^- and present in higher concentration).

Under the buffered conditions one might expect linear first-order plots for the disappearance of acetate ion (which was the assay used to follow the kinetics). In fact a straight line was obtained for the solvolysis of 4 but the first-order plot for solvolysis of 5 showed considerable curvature (see Figure 4.3). Remarkably, most of the curvature occurred *early* in the reaction unlike the graphs in Figure 4.1. The solvolysis exhibited no common ion rate depression, as expected if the acetate buffer had effectively suppressed chloride ion return.

Product analysis revealed the cause of this strange behavior. Winstein and Goering found that 5 underwent two competitive reactions: the first was the

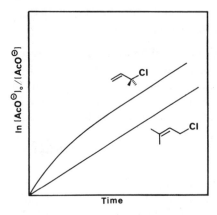

FIGURE 4.3. Solvolyses of 1-chloro-3-methyl-2-butene (**4**) and 3-chloro-3-methyl-1-butene (**5**). The data are plotted in the form that would be appropriate for a first order irreversible reaction.

expected solvolysis but the second was a rearrangement to give **4**. The early high rate of acetate ion removal was due to the solvolysis of **5** but the later, slower reaction was really the solvolysis of **4**, produced from **5** by rearrangement. The crucial feature of this observation was that **4** could not have been produced from **5** by way of the carbonium ion **6** since the lack of common ion rate depression

Products

indicated that there was no significant choloride ion return. The possibility of a direct interconversion of **5** and **4** (what we would not call a [1,3] sigmatropic shift) was considered by Winstein and Goering but rejected because the rearrangement was found to occur only in polar media used for solvolysis

Products

reactions. In order to explain their data they postulated the intermediacy of an ion pair. Ion pairs had previously been proposed by Bjerrum and Fuoss to explain the anomalous conductivity of concentrated solutions of strong electrolytes, but this was the first indication that they might play an important mechanistic role in organic reactions. With the introduction of this concept it was possible to extend the Ingold scheme for S_N1 reactions:

$$R-X \rightleftharpoons \underset{\text{ion pair}}{R^+X^-} \rightleftharpoons R^+ + X^- \xrightarrow{\text{Nu:}} \text{Products}$$

The next major addition to this scheme came in 1958, again from the work of Winstein's group.[7] The substrate under scrutiny was optically active brosylate (p-bromobenzenesulfonate) 7 (see Figure 4.4). The progress of its solvolysis could be followed by two independent methods—polarimetric measurement of the racemization of the reaction mixture, with phenomenological rate constant k_α, and titrimetric determination of the formation of p-bromobenzenesulfonic acid, with rate constant k_t. If the mechanism depicted in Figure 4.4 were correct then one could equate the phenomenological rate constant k_α with the mechanistic rate constant k_1. The titrimetric rate constant k_t can be evaluated by applying the steady-state approximation to ion pair 8 and carbonium ion 9. The result is

$$\frac{d[\text{H}^+]}{dt} = \frac{k_1 k_2 k_3}{k_3(k_{-1} + k_2) + k_{-1}k_{-2}[\text{X}^-]} \cdot [\text{RX}]$$

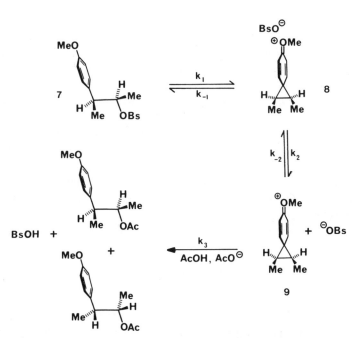

FIGURE 4.4. Solvolysis of optically active 7. The first-formed intermediate is an achiral ion pair in this mechanistic proposal.

Hence,

$$k_t = \frac{k_1 k_2 k_3}{k_3(k_{-1} + k_2) + k_{-1} k_{-2}[\text{X}^-]}$$

From this expression one might expect to see common ion rate depression on k_t. In fact none was observed, indicating that, for the mechanism to be correct, the mechanistic rate constants must have values that make the term containing $[\text{X}^-]$ in the denominator negligible, that is, $k_3(k_{-1} + k_2) \gg k_{-1} k_{-2}[\text{X}^-]$. Hence,

$$k_t = \frac{k_1 k_2}{k_{-1} + k_2} \tag{13}$$

The mechanism represented by Figure 4.4 and Eq. (13) is subject to experimental test. The test takes the form of a salt effect study and can be summarized as follows:

$$k_\alpha = k_1$$

Hence from Eq. (13),

$$\frac{k_\alpha}{k_t} = \frac{k_{-1}}{k_2} + 1$$

The effect of increasing the ionic strength of the medium should be to increase k_2 and decrease k_{-1}. Hence for some non-nucleophilic ionic species such as LiClO_4, one would expect that $k_\alpha/k_t \rightarrow 1$ as $[\text{LiClO}_4] \rightarrow \infty$.

The experimental observation was not at all in accord with this expectation (see Figure 4.5). Not only did k_t fail to approach k_α at high LiClO_4 concentrations but, in addition, k_t exhibited a strong, nonlinear dependence at low LiClO_4 concentrations. The latter was termed the "special salt effect" by Winstein.

In order to explain the data Winstein proposed the existence of two types of ion pair, the so-called contact (or intimate) ion pair and the solvent-separated ion pair, written, respectively, as R^+X^- and $\text{R}^+ \parallel \text{X}^-$.

The generalized solvolysis scheme thus becomes

$$\text{RX} \underset{k_{-1}}{\overset{k_1}{\rightleftharpoons}} \text{R}^+\text{X}^- \underset{k_{-2}}{\overset{k_2}{\rightleftharpoons}} \text{R}^+\parallel\text{X}^- \underset{k_{-3}}{\overset{k_3}{\rightleftharpoons}} \text{R}^+ + \text{X}^- \overset{k_4}{\longrightarrow} \text{products}$$

The rate constants k_{-1}, k_{-2}, and k_{-3} define processes that Winstein has called internal return, external ion pair return, and external ion return, respectively.

The phenomenon known as the special salt effect was interpreted by Winstein

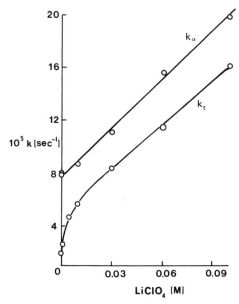

FIGURE 4.5. The special salt effect of LiClO₄ on the solvolysis of **7**. Reprinted with permission from S. Winstein and G. C. Robinson, *J. Am. Chem. Soc.*, **80**, 169 (1958). Copyright 1958, American Chemical Society.

to be due to interception of the solvent-separated ion pair by perchlorate ions:

$$R^+ \| X^- + ClO_4^- \rightleftharpoons R^+ \| ClO_4^- + X^-$$
$$\downarrow$$
$$R^+ + ClO_4^-$$

The non-nucleophilic counterion ClO_4^- was presumed to suppress external ion pair return, thereby increasing the rate of dissociation to free ions and hence the rate of product formation. This explanation did lead to an experimentally testable prediction, namely that addition of X^- to a system experiencing the special salt effect should drive the equilibrium between solvent-separated ion pairs back in favor of $R^+ \| X^-$, thereby reducing the magnitude of k_t. The existence of this so-called "induced common ion rate depression" was confirmed experimentally,[8] strengthening Winstein's proposition considerably.

In keeping with Occam's razor, one might question whether it is really necessary to have two types of ion pair in addition to the dissociated ions as intermediates in a solvolysis reaction. In particular, if solvent-separated ion pairs can be intercepted by perchlorate could they not also be intercepted by other nucleophiles to give products? If so, would one need to invoke the formation of free carbonium ions any more?

It turns out to be easy to answer the second question. The solution lies in working out the rate law for the generalized solvolysis scheme shown in Eq. (14).

$$RX \underset{k_{-1}}{\overset{k_1}{\rightleftharpoons}} R^+X^- \underset{k_{-2}}{\overset{k_2}{\rightleftharpoons}} R^+\|X^- \underset{k_{-3}}{\overset{k_3}{\rightleftharpoons}} R^+ + X^-$$

$$\downarrow k_4 \qquad\qquad \downarrow k_5 \tag{14}$$

$$\text{Products} \quad\ \text{Products}$$

The somewhat ugly result is

$$\frac{d[H^+]}{dt} = \frac{k_1 k_2 \{k_4 + k_3 k_5/(k_5 + k_{-3}[X^-])\}[RX]}{(k_{-1} + k_2)\{k_4 + k_3 k_5/(k_5 + k_{-3}[X^-])\} + k_{-1}k_{-2}} \tag{15}$$

One can simulate a situation in which free carbonium ions are never produced by making $k_4 \gg k_3$. The result is

$$\text{Case I:} \quad \frac{d[H^+]}{dt} = \frac{k_1 k_2 k_4 [RX]}{(k_{-1} + k_2)k_4 + k_{-1}k_{-2}}$$

Since this rate equation contains no term in $[X^-]$, it would not be possible to see common ion rate depression. Thus observation of common ion rate depression requires the existence of free carbonium ions. Ingold's solvolysis of 4,4'-dimethylbenzhydryl chloride provides an example.

But this was our second question, the first still remains. Can one form products from solvent-separated ion pairs? In view of the discussion above, one might suspect that failure to see common ion rate depression would be diagnostic of a reaction in which products were formed at the solvent-separated ion pair stage. Sadly, this is not correct. In addition to Case I, considered above, one can construct three more situations in which common ion rate depression would not be observed:

If $k_5 \gg k_{-3}[X^-]$

$$\text{Case II:} \quad \frac{d[H^+]}{dt} = \frac{k_1 k_2 (k_4 + k_3)[RX]}{(k_{-1} + k_2)(k_4 + k_3) + k_{-1}k_{-2}}$$

If $k_{-3}[X^-] \gg k_5$ and $k_3 \simeq k_4$

$$\text{Case III:} \quad \frac{d[H^+]}{dt} = \frac{k_1 k_2 k_4 [RX]}{(k_{-1} + k_2)k_4 + k_{-1}k_{-2}}$$

If $k_{-1} \simeq 0$ or $k_{-2} \simeq 0$

$$\text{Case IV:} \quad \frac{d[H^+]}{dt} = \frac{k_1 k_2 [RX]}{k_{-1} + k_2}$$

Only Cases I and III would *require* that products be formed from the solvent-separated ion pair. The other two cases would permit it, but it could not be

TABLE 4-2. Substrates Used in the Search for product Formation from Solvent-Separated Ion Pairs

Substrate	CIRD[a]	X⁻/X*	
		Exchange	Interpretation
	Yes	Yes	Free R^+
	No	No	Case IV
7	No	Yes	Case I, II, or III

[a]Common ion rate depression.

proven kinetically. The problem, then, is first to find a solvolysis that does not exhibit common ion rate depression and then to attempt to rule out Cases II and IV as explanations for its absence. The task is made somewhat simpler when one recognizes that Case IV would prohibit incorporation of labeled X^- (designated X^{*-}) into the starting material.

Typical substrates examined by Winstein and coworkers are shown in Table 4.2. Of these three compounds, only brosylate 7 merited further consideration. For this compound the problem was to distinguish Case I or III from Case II. Winstein was able to rule out Case II by running the solvolysis in the presence of azide (N_3^-). The mathematical condition leading to Case II, $k_5 \gg k_{-3}[X^-]$, implies the existence of a very unselective carbonium ion that is influenced only by the relative concentrations of nucleophiles not their intrinsic nucleophilicity. However, when N_3^- was added to the reaction the products were found to be almost exclusively organic azides rather than the acetates that had been formed previously. This high selectivity allows one to rule out Case II, leaving Cases I and III as the only viable candidates. Both Cases I and III require product formation from a solvent-separated ion pair.

The details of ion pair formation and reaction continue to be elucidated but the fundamental concept owes its existence to the kinetic studies of Winstein.

4.5. THE KINETICS OF UNIMOLECULAR ARRAYS

A reaction mechanism does not have to be very complex before the corresponding integrated rate equations become very cumbersome or even impossible to
write in closed form. In many such cases one is forced to resort to numerical
integration, with the attendant problems that we will see in Section 4.6. In the
special case of unimolecular arrays, however, even quite complex examples can
have rate law that are amenable to analytical integration. In this section we will
discuss the mathematical technique that makes this possible. It is especially
powerful when applied to interconversions between enantiomers and label
isomers. The approach described here is related to ideas published by Matsen,[9]
Ritchie,[10] and Goldstein[11] but differs in some respects from all of these.

4.5.1. Mathematical Development

The ultimate goal of any kinetic analysis is to relate mechanistic (hypothetical)
rate constants to phenomenological (observable) rate constants. In this discussion we will represent the mechanistic rate constants by the symbol k_{ij} where the
subscripts imply that the rate constant is for the elementary step that converts
species i to species j.

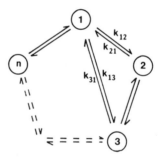

The phenomenological rate constants will be represented by the symbol λ_i.
These symbols will not be retained in the discussion of specific examples but the
distinction between mechanistic and phenomenological rate constants will be
made explicit.

Let A_i represent the concentration of compound i. The differential rate
equation for disappearance of compound 1 can thus be written:

$$\frac{-dA_1}{dt} = (k_{12} + k_{13} + \cdots k_{1n})A_1 - k_{21}A_2 - k_{31}A_3 - \cdots k_{n1}A_n$$

In general,

$$\frac{-dA_i}{dt} = (k_{i1} + k_{i2} + \cdots k_{in})A_i - k_{1i}A_1 - k_{2i}A_2 - \cdots k_{ni}A_n$$

Define

$$K_{ii} = \sum_{l \neq i} k_{il}$$

$$K_{ij} = -k_{ji} \qquad (j \neq i)$$

The complete set of differential equations can thus be summarized in the equation:

$$\frac{-dA_i}{dt} = \sum_{j=1}^{n} K_{ij}A_j \qquad i = 1,2,3 \ldots n$$

In matrix notation this can be written:

$$\frac{-d}{dt}\begin{bmatrix} A_1 \\ A_2 \\ \vdots \\ A_n \end{bmatrix} = \begin{bmatrix} K_{11} & K_{12} & \cdots & K_{1n} \\ K_{21} & K_{22} & & \vdots \\ \vdots & & & \vdots \\ K_{n1} & \cdots & \cdots & K_{nn} \end{bmatrix}\begin{bmatrix} A_1 \\ A_2 \\ \vdots \\ A_n \end{bmatrix}$$

$$\qquad \mathbf{A} \qquad\qquad\qquad \mathbf{K}$$

$$\frac{-d\mathbf{A}}{dt} = \mathbf{K} \cdot \mathbf{A} \qquad\qquad (16)$$

Note the similarity of Eq. (16) to the differential equation for a single, irreversible, first-order reaction (Section 4.1). The similarity persists in the integrated form, with matrices replacing the more familiar scalar variables. Thus a particular solution is

$$\mathbf{A} = \mathbf{B}e^{-\mathbf{\Lambda}t}$$

The general solution is a linear combination of particular solutions:

$$\mathbf{A} = \mathbf{B}e^{-\mathbf{\Lambda}t}\mathbf{Q}^{\circ}$$

where $\mathbf{Q}^{\circ}$ is a column vector of linear combination coefficients. One can evaluate these coefficients by considering a boundary condition:

At $t = 0$: $\mathbf{A} = \mathbf{A}^{\circ}$; $\mathbf{\Lambda}t = \mathbf{0}$ (null matrix); $e^{-\mathbf{\Lambda}t} = \mathbf{I}$ (unit matrix):

$$\mathbf{A}^{\circ} = \mathbf{B} \cdot \mathbf{Q}^{\circ}$$

$$\mathbf{Q}^{\circ} = \mathbf{B}^{-1} \cdot \mathbf{A}^{\circ}$$

Hence,

$$A = Be^{-\Lambda t}B^{-1}A^{\circ} \qquad (17)$$

Hence,

The nature of B and Λ can be deduced by differentiation of Eq. (17) and resubstitution into Eq. (16):

$$\frac{-dA}{dt} = B\Lambda e^{-\Lambda t}B^{-1}A^{\circ}$$

$$= KA$$

$$= KBe^{-\Lambda t}B^{-1}A^{\circ}$$

Hence,

$$B\Lambda = KB$$

$$\Lambda = B^{-1}KB$$

Thus it can be seen that Λ and B are, respectively, the diagonal matrix of eigenvalues and the matrix of eigenvectors derived from the rate constant matrix K (see Appendix 1). Assuming that Λ and B can be found (more about that later), substitution into Eq. (17) provides the closed form of the integrated rate equations. Note, however, that implicit in this substitution is a matrix inversion to find B^{-1}. This inversion is often quite tedious, but it can be avoided by doing a little extra work before performing the integration.

The trick is to make K into a symmetric matrix (which we will call K_s). If the corresponding matrix of eigenvectors is called B_s then it is always true that $B_s^{-1} = \widetilde{B}_s$, where $\widetilde{B}_s$ is just the transpose of B_s. The symmetrization of K is achieved simply by replacing each corresponding pair of off-diagonal elements, $-k_{ji}$ and $-k_{ij}$, by $-\sqrt{(k_{ji}k_{ij})}$. This procedure can be justified by consideration of the principle of detailed balance as shown in Appendix 2. For the present purposes we need merely note that the symmetrization procedure is equivalent to carrying out the multiplication:

$$K_s = S^{-1/2}KS^{1/2}$$

where

$$S^{1/2} = \begin{bmatrix} \sqrt{A_1^{\infty}} & 0 \cdots \cdots \cdots & 0 \\ 0 & \sqrt{A_2^{\infty}} & \\ \vdots & & \vdots \\ 0 & \cdots \cdots \cdots \cdots & \sqrt{A_n^{\infty}} \end{bmatrix}$$

$$S^{-1/2} = \begin{bmatrix} 1/\sqrt{A_1^\infty} & 0 \cdots \cdots \cdots & 0 \\ 0 & 1/\sqrt{A_2^\infty} & 0 \\ \vdots & & \vdots \\ 0 & \cdots \cdots \cdots & 1/\sqrt{A_n^\infty} \end{bmatrix}$$

Since $\mathbf{K}$ and $\mathbf{K}_s$ are similar matrices, they must have the same eigenvalues but different eigenvectors:

$$\Lambda = \mathbf{B}^{-1}\mathbf{K}\mathbf{B}$$

$$= \mathbf{B}_s^{-1}\mathbf{K}_s\mathbf{B}_s$$

$$= \widetilde{\mathbf{B}}_s\mathbf{K}_s\mathbf{B}_s$$

The only remaining problem is that the symmetrized rate constant matrix does not fit the original set of differential equations:

$$\frac{-d\mathbf{A}}{dt} \neq \mathbf{K}_s\mathbf{A}$$

We need to find some $\mathbf{A}_s$ such that

$$\frac{-d\mathbf{A}_s}{dt} = \mathbf{K}_s\mathbf{A}_s$$

This turns out to be easy if one simply premultiplies Eq. (16) by $\mathbf{S}^{-1/2}$:

$$\frac{-d\mathbf{A}}{dt} = \mathbf{K}\mathbf{A}$$

$$-\mathbf{S}^{1/2}\frac{d\mathbf{A}}{dt} = \mathbf{S}^{-1/2}\mathbf{K}\mathbf{A}$$

$$\frac{-d\{\mathbf{S}^{-1/2}\mathbf{A}\}}{dt} = \mathbf{S}^{-1/2}\mathbf{K}\mathbf{S}^{1/2}\mathbf{S}^{-1/2}\mathbf{A}$$

$$= \mathbf{K}_s\{\mathbf{S}^{-1/2}\mathbf{A}\}$$

Hence,

$$\mathbf{A}_s = \mathbf{S}^{-1/2}\mathbf{A}$$

This transformed concentration vector can now be incorporated into the final integrated rate equation:

$$\mathbf{A}_s = \mathbf{B}_s e^{-\mathbf{\Lambda' \widetilde{B}_s A_s^\circ}} \qquad (18)$$

in which, as desired, the inversion of the matrix of eigenvectors has been eliminated. Equation (18) will form the basis for the rest of our discussion in Section 4.5. We begin with an illustration of its application to the stereomutation of 1-phenylcyclopropane-2-d.

4.5.2. Stereomutation of 1-Phenylcyclopropane-2-d

The stereomutation (cis-trans isomerization plus optical isomerization) of cyclopropanes might appear to be a very simple reaction for which it would be difficult to write many mechanisms. In fact three mechanisms have been proposed and there has been considerable controversy about which (if any) is correct. The three mechanisms are:

A. Single Methylene Rotation (Smith[12]).   This mechanism begins by undergoing a reaction similar to retro-cycloaddition of a carbene to an olefin. Instead of proceeding to complete dissociation, however, the "carbene–olefin complex" undergoes an internal rotation and then collapses back to an isomerized cyclopropane.

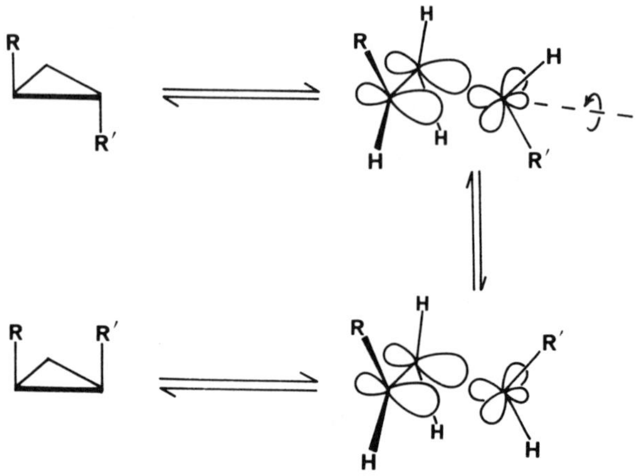

B. Concerted Double Methylene Rotation (Hoffmann[13]).   On the basis of extended Hückel calculations, Hoffman proposed that thermolysis of cyclopropane proceeds by conrotatory cleavage of a C—C bond followed by conrotatory closure of the trimethylene biradical.

C. Random Pairwise Methylene Rotation (Benson[14]).   Benson used thermochemical estimates to calculate the heat of formation of the trimethylene

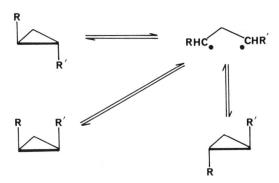

biradical. He concluded that it should lie approximately 9 kcal/mol below the transition state for stereomutation. A potential energy well this deep would allow the biradical to undergo many rotations about the C—C bonds before reclosing to cyclopropane.

Assuming that these rotations were uncorrelated, the effect would be to randomize the stereochemistry at two sites each time a C—C bond of cyclopropane was broken. (Of course if neither R nor R' was an isotopic label there would be transition states of different energy leading to the *cis* and *trans* cyclopropanes. In this sense the stereochemistry would not be strictly randomized. For the present example, 1-phenylcyclopropane-2-*d*, no such qualification exists provided one considers equilibrium isotope effects to be negligible.)

The important distinguishing feature for each of these mechanisms is the number and nature of the isomers that can be produced in a single turnover. Mechanism A can *cis-trans* isomerize in a single step but requires two steps to interconvert enantiomers. Mechanism B can interconvert enantiomers by breaking the HRC—CR'H bond but must break a H_2C—CRH bond to cause *cis-trans* isomerization. Mechanism C allows formation of both *cis-trans* isomers and optical isomers from a single cleavage of the HRC—CR'H bond. These differences form the basis for an experimental differentiation among the mechanisms.

In order to keep the present example as simple as possible we will not explicitly include the Benson biradical, since to do so requires some further mathematical

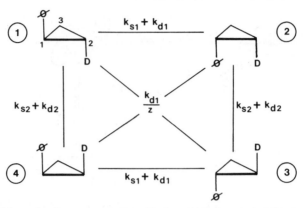

FIGURE 4.6. Scheme for the stereomutation of 1-phenylcyclopropane-2-*d*. See text for definition of the mechanistic rate constants.

manipulation. The biradical will be included when we discuss cyclopropane-1,2-d_2. The mechanistic scheme for interconversion of the isomers of 1-phenyl-cyclopropane-2-*d* is shown in Figure 4.6. The mechanistic rate constants are defined as follows:

k_{s1} = single rotation at C1
k_{s2} = single rotation at C2
k_{d1} = double rotation at C1, C3
k_{d2} = double rotation at C2, C3
z = secondary deuterium isotope effect

As shown in Figure 4.6, the mechanistic rate constant for double rotation at C1, C2 should have the same magnitude as that for double rotation at C1, C3 except for a secondary kinetic isotope effect, z. Isotope effects of this kind will be discussed in detail in Chapter 5. For now we can just treat z as some correction factor. Implicit in Figure 4.6 is the assumption that all equilibrium isotope effects are negligible, that is, that all equilibrium constants are unity. This assumption is almost certainly valid within the limits of precision to be expected for a typical kinetics experiment.

The first step toward a solution to the kinetic problem represented by Figure 4.6 is creation of the rate constant matrix **K**. Remembering the definitions of K_{ii} and K_{ij} in Section 4.5.1, we can write

$$\mathbf{K} = \begin{bmatrix} a+b+c & -a & -c & -b \\ -a & a+b+c & -b & -c \\ -c & -b & a+b+c & -a \\ -b & -c & -a & a+b+c \end{bmatrix}$$

where $a = k_{s1} + k_{d1}$, $b = k_{s2} + k_{d2}$, $c = k_{d1}/z$.

The next step would normally be to symmetrize this matrix except that, in the present example, $\mathbf{K}$ is already symmetric because all equilibrium constants were unity. Hence,

$$\mathbf{K}_s = \mathbf{K}$$

In order to find the eigenvalues and eigenvectors of $\mathbf{K}_s$ one could follow the procedure detailed in Appendix 1. In a future example we will have to do this but, again, the rearrangement of 1-phenylcyclopropane-2-d is made simple by its symmetry. Appendix 3 lists some generalized solutions to kinetic problems that possess certain symmetry properties. The present example will be recognized as Case II of the four component class. The eigenvalues and eigenvector matrix are thus:

$$\lambda_1 = 0$$

$$\lambda_2 = 2[k_{s1} + k_{d1}(1 + 1/z)] \tag{19}$$

$$\lambda_3 = 2(k_{s1} + k_{d1} + k_{s2} + k_{d2}) \tag{20}$$

$$\lambda_4 = 2(k_{s2} + k_{d1}/z + k_{d2}) \tag{21}$$

$$\mathbf{B}_s = \frac{1}{2}\begin{bmatrix} 1 & 1 & 1 & 1 \\ 1 & -1 & -1 & 1 \\ 1 & -1 & 1 & -1 \\ 1 & 1 & -1 & -1 \end{bmatrix}$$

For a closed system one of the eigenvalues (arbitrarily called λ_1) is always zero. The reason for this will be discussed in Section 4.5.4.

Since the original rate constant matrix, $\mathbf{K}$, for this problem was already symmetric, the matrix $\mathbf{S}^{-1/2}$ is just some constant times the identity. Hence,

$$\mathbf{A}_s = \mathbf{A}$$

We now have all of the matrices necessary for substitution into Eq. (18). (See Appendix 1 for the evaluation of $e^{-\boldsymbol{\Lambda}t}$.) The final integrated rate equations take the form:

$$A_1 = (1/4)(S_1 + S_2e^{-\lambda_2 t} + S_3e^{-\lambda_3 t} + S_4e^{-\lambda_4 t})$$

$$A_2 = (1/4)(S_1 - S_2e^{-\lambda_2 t} - S_3e^{-\lambda_3 t} + S_4e^{-\lambda_4 t})$$

$$A_3 = (1/4)(S_1 - S_2e^{-\lambda_2 t} + S_3e^{-\lambda_3 t} - S_4e^{-\lambda_4 t})$$

$$A_4 = (1/4)(S_1 + S_2e^{-\lambda_2 t} - S_3e^{-\lambda_3 t} - S_4e^{-\lambda_4 t})$$

$$S_1 = A_1^\circ + A_2^\circ + A_3^\circ + A_4^\circ$$

$$S_2 = A_1^\circ - A_2^\circ - A_3^\circ + A_4^\circ$$

$$S_3 = A_1^\circ - A_2^\circ + A_3^\circ - A_4^\circ$$

$$S_4 = A_1^\circ + A_2^\circ - A_3^\circ - A_4^\circ$$

Networks of unimolecular reactions always have integrated rate equations that can be written as sums of exponentials:

$$A_i = \sum_{j=1}^{n} c_{ij} e^{-\lambda_j t}$$

The λ_j are the experimentally determinable rate constants that we now recognize to be the eigenvalues of the mechanistic rate constant matrix. Our analysis has thus allowed us to relate phenomenological and mechanistic rate constants, which was our stated goal. One more step is necessary, however, before solution can be considered complete. That is to find out *which particular* phenomenological rate constants correspond to which combinations of mechanistic rate constants. This entails consideration of the actual experiments to be conducted.

For 1-phenylcyclopropane-2-*d* an obvious experiment is to determine the rate constant for one racemic stereoisomer, say the *trans*, to approach equilibrium with the other, the *cis*. In other words one would observe the depletion of $A_1 + A_3$. Let this phenomenological rate constant be called k_i. From the integrated rate equations we find that

$$A_1 + A_3 = (1/2)(A_1 + S_3 e^{-\lambda_3 t})$$

Hence,

$$k_i = \lambda_3$$

A second experiment might be to prepare optically active **1** and to monitor for disappearance of optical activity upon pyrolysis. A little reflection shows that there is a complication with this experiment, however. The optical activity at any time is given by

$$\alpha_{\mathrm{obs}} = [\alpha]_t^\circ (A_1 - A_3) + [\alpha]_c^\circ (A_4 - A_2)$$

$$[\alpha]_t^\circ = \text{specific rotation for pure } \mathbf{1}$$

$$[\alpha]_c^\circ = \text{specific rotation for pure } \mathbf{4}$$

Note that to get the signs right in this equation one must, by some experiment, relate the absolute configurations of *cis* and *trans* isomers. The time evolution of the optical activity is derived by substituting for the concentration terms using the integrated rate equations. The result is

$$\alpha_{obs} = (1/2)([\alpha]_t^\circ + [\alpha]_c^\circ)S_2 e^{-\lambda_2 t} + (1/2)([\alpha]_t^\circ - [\alpha]_c^\circ)S_4 e^{-\lambda_4 t}$$

Now we see the complication. The expression for the time evolution of α_{obs} contains two exponential terms and so it would not be possible to determine a first-order rate constant except in the unlikely event that *cis* and *trans* isomers had identical specific rotations, in which case the second term would vanish.

One could make the second term disappear deliberately by making $S_4 = 0$. This could be achieved if $A_1^\circ = A_3^\circ$; $A_2^\circ = A_4^\circ = 0$. In other words, one would make up a mixture containing equal proportions of *cis* and *trans* isomers having the same absolute configuration at C1 and observe the decline of optical activity upon pyrolysis. This process would have a first-order time dependence and the phenomenological rate constant, k_α, could be equated with λ_2.

The eigenvalue λ_4 could be found by returning to the experiment in which pure 1 was pyrolyzed and substituting in the known value for λ_2:

$$\lambda_4 = \frac{1}{t} \left\{ \frac{(1/2)([\alpha]_t^\circ - [\alpha]_c^\circ)A_1^\circ}{\alpha_{obs} - (1/2)([\alpha]_t^\circ + [\alpha]_c^\circ)A_1^\circ e^{-\lambda_2 t}} \right\}$$

Finally, knowing that the three nonzero eigenvalues can be determined experimentally, we can reformulate Eqs. (19)–(21) to assign values to the mechanistic rate constants:

$$k_{s1} = (1/4)\{\lambda_2(1 - z) + (\lambda_3 - \lambda_4)(1 + z)\}$$

$$k_{d1} = (z/4)\{\lambda_2 - \lambda_3 + \lambda_4\}$$

$$k_{s2} + k_{d2} = (1/4)\{\lambda_3 + \lambda_4 - \lambda_2\}$$

Note that the problem is underdetermined (there are more mechanistic rate constants than phenomenological rate constants) and so k_{s2} and k_{d2} cannot be separated.

The experiments described above were actually carried out by Berson and Pedersen.[15] Using an assumed value of $z = 1.10$ (see Chapter 5 for justification) they found:

$$k_{s1} = 0.0$$

$$k_{d1} = 9.6 \times 10^{-6} \ \text{sec}^{-1}$$

$$k_{s2} + k_{d2} = 1.9 \times 10^{-6} \ \text{sec}^{-1}$$

This corresponds to 83% of the reaction being a concerted double rotation involving C1 and 17% being either a double rotation at C2, C3 or a single rotation at C2. For this particular cyclopropane, then, the Hoffman mechanism seems to be dominating. Other cyclopropanes seem to show higher proportions of single rotation (or random rotation) for reasons that are not clear.[16]

4.5.3. Treatment of Steady-State Intermediates—Stereomutation of Cyclopropane-1,2-d$_2$

Consider the reaction scheme:

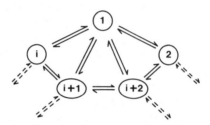

The species **1** to **i** have measurable concentrations at some time during the reaction. Species **i + 1** to **n** are reactive intermediates that would normally be dealt with in a kinetic analysis by applying the steady-state approximation to their concentrations.

The differential rate equations can be written in block matrix form:

$$\frac{-d}{dt}\begin{bmatrix} \mathbf{A}_\alpha \\ \mathbf{A}_\beta \end{bmatrix} = \begin{bmatrix} \mathbf{K}_{\alpha\alpha} & \mathbf{K}_{\alpha\beta} \\ \mathbf{K}_{\beta\alpha} & \mathbf{K}_{\beta\beta} \end{bmatrix} \begin{bmatrix} \mathbf{A}_\alpha \\ \mathbf{A}_\beta \end{bmatrix}$$

where $\mathbf{A}_\alpha$ = vector of concentrations A_1 to A_i
   $\mathbf{A}_\beta$ = vector of concentrations A_{i+1} to A_n.
   $\mathbf{K}_{\alpha\alpha}$ = rate constants interconverting species **1** to **i**.
   $\mathbf{K}_{\beta\beta}$ = rate constants interconverting species **i + 1** to **n**.
   $\mathbf{K}_{\alpha\beta}$, $\mathbf{K}_{\beta\alpha}$ = rate constants between (**1** to **i**) and (**i + 1** to **n**).

Expanding

$$\frac{-d}{dt} \mathbf{A}_\alpha = \mathbf{K}_{\alpha\alpha}\mathbf{A}_\alpha + \mathbf{K}_{\alpha\beta}\mathbf{A}_\beta$$

$$\frac{-d}{dt} \mathbf{A}_\beta = \mathbf{K}_{\beta\alpha}\mathbf{A}_\alpha + \mathbf{K}_{\beta\beta}\mathbf{A}_\beta$$

$$= \mathbf{0} \text{ (steady-state approximation)}$$

$$\mathbf{A}_\beta = -\mathbf{K}_{\beta\beta}^{-1}\mathbf{K}_{\beta\alpha}\mathbf{K}_\alpha$$

$$\frac{-d}{dt} \mathbf{A}_\alpha = \mathbf{K}_{\alpha\alpha}\mathbf{A}_\alpha - \mathbf{K}_{\alpha\beta}\mathbf{K}_{\beta\beta}^{-1}\mathbf{K}_{\beta\alpha}\mathbf{A}_\alpha$$

$$= \{ K_{\alpha\alpha} - K_{\alpha\beta} K_{\beta\beta}^{-1} \ K_{\beta\alpha} \} \ A_{\alpha}$$

$$= K_{\alpha\alpha}' A_{\alpha} \tag{22}$$

$K_{\alpha\alpha}'$ is a rate constant matrix of dimensions $i \times i$.

An illustration of how one applies Eq. (22) is provided by the stereomutation of optically active cyclopropane-1,2-d_2. This time we will explicitly consider the biradical intermediate proposed by Benson (see Figure 4.7). The stereochemistry at the radical carbons is not shown because, consistent with Benson's hypothesis, we assume that stereochemical information is lost at these centers as soon as the biradical is formed (or at least at a rate much greater than the rate of ring closure). The mechanistic rate constants shown in Figure 4.7 have the following definitions.

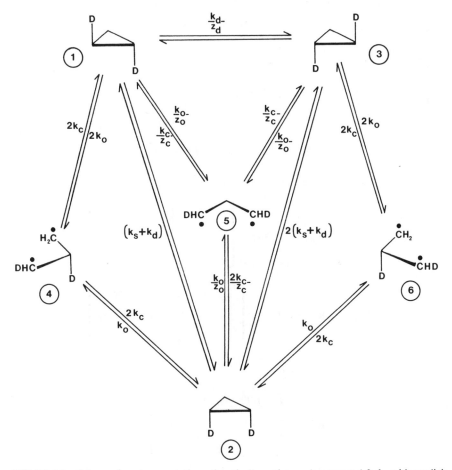

FIGURE 4.7. Scheme for stereomutation of optically active cyclopropane-1,2-d_2, with explicit inclusion of Benson's biradicals. The rate constants for conversion of 1 to 2, 3 to 1, and 3 to 2 are defined by symmetry. They are $2(k_s + k_d)$, (k_d/z_d), and $(k_s + k_d)$, respectively.

k_s = rate constant for single CHD rotation (Smith mechanism).

k_d = rate constant for correlated rotation of CH_2 and CHD (Hoffmann mechanism).

k_o = rate constant for opening to biradical (Benson mechanism).

k_c = rate constant for closure of biradical intermediate.

The secondary kinetic isotope effects on double rotation, opening to the biradical, and closure of the biradical are designated z_d, z_o, and z_c, respectively. In each case the isotope effect refers to the relative magnitude of the rate constants for reaction at CHD, CH_2 vs. CHD, CHD. No isotope effect for single rotation need be specified since this mechanism cannot be detected when a CH_2 group rotates.

The rate constant matrix for the scheme in Figure 4.7 is:

$$
\left[
\begin{array}{ccc|ccc}
\Sigma\,k + k_d/z_d & -(k_s + k_d) & -k_d/z_d & -2k_c & -k_c/z_c & 0 \\
-2(k_s + k_d) & \Sigma\,k & -2(k_s + k_d) & -2k_c & -2k_c/z_c & -2k_c \\
-k_d/z_d & -(k_s + k_d) & \Sigma\,k + k_d/z_d & 0 & -k_c/z_c & -2k_c \\
\hline
-2k_o & -k_o & 0 & 4k_c & 0 & 0 \\
-k_o/z_o & -k_o/z_o & -k_o/z_o & 0 & 4k_c/z_c & 0 \\
0 & -k_o & -2k_o & 0 & 0 & 4k_c
\end{array}
\right]
$$

$$ \Sigma\,k = 2k_s + k_d + k_o(2 + 1/z_o) $$

Applying the steady-state approximation:

$$
\mathbf{K}_{\beta\beta}^{-1} = 1/(4k_c)
\begin{bmatrix}
1 & 0 & 0 \\
0 & z_c & 0 \\
0 & 0 & 1
\end{bmatrix}
$$

$$
\mathbf{K}_{\alpha\beta}\mathbf{K}_{\beta\beta}^{-1}\mathbf{K}_{\beta\alpha} = (k_o/2)
\begin{bmatrix}
2 + x & 1 + x & x \\
2 + 2x & 2 + 2x & 2 + 2x \\
x & 1 + x & 2 + x
\end{bmatrix}
$$

$$ x = 1/(2z_c) $$

$$
\mathbf{K}'_{\alpha\alpha} =
\begin{bmatrix}
a + b & -a/2 & -b \\
-a & a & -a \\
-b & -a/2 & a + b
\end{bmatrix}
$$

$$ a = k_o(1 + 1/2z_o) + 2(k_s + k_d) $$

$$ b = k_o/(4z_o) + k_d/z_d $$

Symmetrizing

$$(\mathbf{K'_{\alpha\alpha}})_s = \begin{bmatrix} a+b & -a/\sqrt{2} & -b \\ -a/\sqrt{2} & a & -a/\sqrt{2} \\ -b & -a/\sqrt{2} & a+b \end{bmatrix}$$

Comparison of this result with the three component case in Appendix 3 shows that they are identical if $n = 1/2$. The eigenvalues and eigenvector matrix are thus:

$$\lambda_1 = 0$$

$$\lambda_2 = a + 2b$$

$$= k_o(1 + 1/z_o) + 2k_s + 2k_d(1 + 1/z_d)$$

$$\lambda_3 = 2a$$

$$= k_o(2 + 1/z_o) + 4k_s + 4k_d$$

$$\mathbf{B}_s = \begin{bmatrix} 1/2 & 1/\sqrt{2} & 1/2 \\ 1/\sqrt{2} & 0 & -1/\sqrt{2} \\ 1/2 & -1/\sqrt{2} & 1/2 \end{bmatrix}$$

And the symmetrization matrix is

$$\mathbf{S}^{-1/2} = \begin{bmatrix} 2 & 0 & 0 \\ 0 & \sqrt{2} & 0 \\ 0 & 0 & 2 \end{bmatrix}$$

Substitution into Eq. (18) gives the final integrated rate equations:

$$A_1 = (1/4)(S_1 + 2S_2e^{-\lambda_2 t} + S_3e^{-\lambda_3 t})$$

$$A_2 = (1/2)(S_1 \qquad\qquad - S_3e^{-\lambda_3 t})$$

$$A_3 = (1/4)(S_1 - 2S_2e^{-\lambda_2 t} + S_3e^{-\lambda_3 t})$$

$$S_1 = A_1^o + A_2^o + A_3^o$$

$$S_2 = A_1^o \qquad\quad - A_3^o$$

$$S_3 = A_1^o - A_2^o + A_3^o$$

As in the previous example, the one remaining problem is to identify which eigenvalues correspond to which phenomenological rate constants.

The experiment in which racemic cyclopropane-*trans*-1,2-d_2 is allowed to reach equilibrium with the *cis* isomer will be a first-order process having a phenomenological rate constant that we can call k_i. The experiment consists of measuring the rate of depletion of $A_1 + A_3$. From the integrated rate equations:

$$A_1 + A_3 = (1/2) (S_1 + S_3 e^{-\lambda_{3'}})$$

Hence,

$$k_i = \lambda_3$$

Starting with just one enantiomer of the *trans* isomer, one could measure the rate of loss of optical activity upon pyrolysis. The rate constant for this process can be called k_α. The optical activity at any time is given by

$$\alpha_{obs} = [\alpha]^\circ (A_1 - A_3)$$

$$= [\alpha]^\circ S_2 e^{-\lambda_{2'}}$$

Where $[\alpha]^\circ$ is the specific rotation of optically pure **1**. Hence,

$$k_\alpha = \lambda_2$$

It is now a simple matter to determine the ratio k_i/k_α that is predicted by the three mechanisms. When considering each mechanism in turn to be the exclusive process one simply sets the mechanistic rate constants for the other two to zero:

$$\frac{k_i}{k_\alpha} = \frac{4k_s + 4k_d \qquad\qquad + k_o(2 + 1/z_o)}{2k_s + 2k_d(1 + 1/z_d) + k_o(1 + 1/z_o)}$$

For mechanism A (Smith) $k_d = k_o = 0$, hence,

$$\frac{k_i}{k_\alpha} = 2.0$$

For mechanism B (Hoffmann) $k_s = k_o = 0$, hence:

$$\frac{k_i}{k_\alpha} = \frac{2}{(1 + 1/z_d)}$$

$$= 1.00 \text{ if } z_d = 1.00$$

$$= 1.05 \text{ if } z_d = 1.10$$

For mechanism C (Benson) $k_s = k_d = 0$, hence,

$$\frac{k_i}{k_\alpha} = \frac{(2 + 1/z_o)}{(1 + 1/z_o)}$$

$$= 1.50 \text{ if } z_o = 1.00$$

$$= 1.52 \text{ if } z_o = 1.10$$

Again, the experiments discussed above were carried out by Berson and Pedersen.[15] The experimental result was

$$\frac{k_i}{k_\alpha} = 1.07 \pm 0.04$$

which clearly fits the Hoffmann mechanism better than the other two.

There is some uncertainly associated with the need to use assumed values for the isotope effects and, in fact, Baldwin[17] has suggested that the dominant isotope effect would be of the β type rather than the α type assumed here (see Chapter 5 for explanation of the terminology). This would be equivalent to making $z_d, z_o < 1$, thereby reducing the magnitude of k_i/k_α predicted for both mechanisms B and C. However, in order to bring the ratio predicted by Benson's mechanism down to the observed value one would have to use $z_o = 0.075$. This corresponds to a β isotope effect of 13.3, which is more than 10 times greater than the largest value ever observed.

Even with the uncertainty in isotope effect, then, Berson and Pedersen's experiment seems most consistent with mechanism B.

4.5.4. Determinantal Evaluation of Eigenvalues and Eigenvectors

The two examples that we have discussed so far have turned out to conform to the generalized solutions listed in Appendix 3. Unfortunately this will not always be true. Consider, for example, interconversion of the trigonal–bipyramidal label isomers 1–3. The mechanistic rate constant k_a refers to a process that interchanges one axial and one equatorial ligand, while k_b refers to a process interchanging both axial ligands with two equatorial ligands (e.g., the Berry pseudorotation).

The symmetry numbers (see Appendix 4) of the three isomers are 6, 2, and 1. This tells one immediately that their interconversion will not conform to the three component case in Appendix 3 since it is limited to situations in which one of the equilibrium constants is unity. Under these circumstances one must resort to the determinantal method for evaluation of eigenvalues and eigenvectors described in Appendix 1.

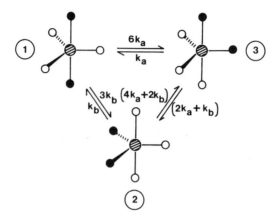

The first step, as always, is to set up the rate constant matrix, **K**.

$$\mathbf{K} = \begin{bmatrix} 6k_a + 3k_b & -k_b & -k_a \\ -3k_b & 4k_a + 3k_b & -(2k_a + k_b) \\ -6k_a & -2(2k_a + k_b) & 3k_a + k_b \end{bmatrix}$$

Since **K** and $\mathbf{K}_s$ are similar matrices they will have the same eigenvalues, and so one could, in principle, find the eigenvalues before or after symmetrization. In fact it is always easier to find them before symmetrization when using the determinant method, for reasons that will become clear in the following discussion. Following the recipe outlined in Appendix 1 we subtract λ from each of the diagonal terms in **K** then write it as a determinant and set it equal to 0:

$$\begin{vmatrix} 6k_a + 3k_b - \lambda & -k_b & -k_a \\ -3k_b & 4k_a + 3k_b - \lambda & -(2k_a + k_b) \\ -6k_a & -2(2k_a + k_b) & 3k_a + k_b - \lambda \end{vmatrix} = 0$$

The eigenvalues are the roots of the polynomial obtained by expansion of this determinant. However, one can save some work by doing a little manipulation on the determinant before expanding it. Remembering that one can add together rows of a determinant without changing its value, we will add rows 2 and 3 to row 1:

$$\begin{vmatrix} -\lambda & -\lambda & -\lambda \\ -3k_b & 4k_a + 3k_b & -(2k_a + k_b) \\ -6k_a & -2(2k_a + k_b) & 3k_a + k_b \end{vmatrix} = 0$$

Note that row 1 now has every element equal to $-\lambda$. This will always occur for a closed system because the diagonal element K_{ii} is the sum of all of the rate constants for the processes destroying species i while the column of elements K_{ji} contains those same rate constants written with negative signs. Consequently the elements in one column of $\mathbf{K}$ must sum to zero or, in the case of the determinant, $-\lambda$. We can now factor out λ from row 1, giving us our first root, $\lambda_1 = 0$. The fact that closed systems always have $\lambda_1 = 0$ is thus recognized as a consequence of the conservation of mass. Subtraction of column 1 from columns 2 and 3 now allows us to reduce the dimensions of the problem by 1. Again, this will always be possible for a closed system provided that one has not symmetrized the rate constant matrix. In the present example the remaining determinant is a 2×2, which can easily be expanded to give a quadratic equation whose roots are found in the normal way.

$$\begin{vmatrix} 1 & 0 & 0 \\ -3k_b & 4k_a + 6k_b - \lambda & -2(k_a - k_b) \\ -6k_a & 2(k_a - k_b) & 9k_a + k_b - \lambda \end{vmatrix} = 0$$

$$\lambda^2 - (13k_a + 7k_b)\lambda + 40k_a^2 + 10k_b^2 + 50k_ak_b = 0$$

$$\lambda_2 = 5k_a + 5k_b$$

$$\lambda_3 = 8k_a + 2k_b$$

The next step is to find the eigenvectors of the symmetrized rate constant matrix $\mathbf{K}_s$. The procedure for doing this is detailed in Appendix 1 and will not be repeated here. The result is

$$\mathbf{B}_s = \begin{bmatrix} 1/\sqrt{10} & 1/\sqrt{2} & \sqrt{6}/\sqrt{15} \\ \sqrt{3}/\sqrt{10} & 1/\sqrt{6} & -2\sqrt{2}/\sqrt{15} \\ \sqrt{6}/\sqrt{10} & -1/\sqrt{3} & 1/\sqrt{15} \end{bmatrix}$$

The effective symmetrization matrix can be determined from the symmetry numbers of the label-isomers (Appendix 4).

$$\mathbf{S}^{-1/2} = \begin{bmatrix} \sqrt{6} & 0 & 0 \\ 0 & \sqrt{2} & 0 \\ 0 & 0 & 1 \end{bmatrix}$$

Now substitution into Eq. (18) gives the integrated rate equations as before.

$$A_1 = (1/10)S_1 + (1/15)S_2 e^{-\lambda_2 t} + (1/6)S_3 e^{-\lambda_3 t}$$

$$A_2 = (3/10)S_1 - (2/15)S_2 e^{-\lambda_2 t} + (1/6)S_3 e^{-\lambda_3 t}$$

$$A_3 = (6/10)S_1 + (1/15)S_2 e^{-\lambda_2 t} - (1/3)S_3 e^{-\lambda_3 t}$$

$$S_1 = A_1^\circ + A_2^\circ + A_3^\circ$$

$$S_1 = 6A_1^\circ - 4A_2^\circ + A_3^\circ$$

$$S_1 = 3A_1^\circ + A_2^\circ - A_3^\circ$$

4.5.5. Treatment of Irreversible Reactions—Rearrangement of *Trans*-1,2-Dipropenylcyclobutane

Detailed studies by Berson and coworkers[18] have shown that pyrolysis of *trans*-1,2-di(*trans*-propenyl)cyclobutane results in the formation of *cis*- and *trans*-3-methyl-4-*trans*-propenylcyclohexene and 3,4-dimethylcycloocta-1,5-diene. In addition, when the starting material is optically active, racemization occurs under the reaction conditions. The dimethylcyclooctadiene presumably arises by *trans* → *cis* isomerization followed by a Cope rearrangement. Independent experiments show that the latter is so fast that the stereochemical isomerization of the *trans*-dipropenylcyclobutane can be treated as effectively irreversible.

The formation of the 3-methyl-4-propenylcyclohexenes represents a formal [1,3] sigmatropic shift. Berson was interested in determining whether this was a true pericyclic process or whether it occurred via a biradical intermediate.

Since a rotationally equilibrated biradical would be achiral, one could hope to attack this problem by using an optically active starting material and looking for optical activity in the products. However, this approach is complicated by the racemization of the starting material. One could have a completely stereospecific rearrangement and yet see loss of optical purity in the products because of this competing side reaction. The problem can be resolved by a kinetic analysis provided that the racemization is not too fast. (If the racemization occurred at a

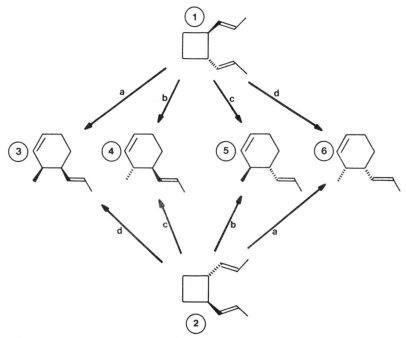

FIGURE 4.8.   Thermal ring expansion of *trans*-1,2-di(*trans*-propenyl)cyclobutane. The rate constants are $a = 2k_{sr}$, $b = 2k_{ar}$, $c = 2k_{si}$, and $d = 2k_{ai}$.

much greater rate than the rearrangement, the products would be completely racemic and no useful information could be gained.) The mechanistic scheme is shown in Figure 4.8.

The mechanistic rate constants have the following definitions.

k_{sr} = rate constant for suprafacial migration with retention of configuration at the migrating carbon.

k_{si} = rate constant for suprafacial migration with inversion of configuration at the migrating carbon.

k_{ar} = rate constant for antarafacial migration with retention of configuration at the migrating carbon.

k_{ai} = rate constant for antarafacial migration with inversion of configuration at the migrating carbon.

The biradical is not explicitly included in this scheme but, as we shall see, it is still possible to deduce something about its existence and properties.

The rate constant matrix for the mechanistic scheme can be set up in the usual way. Note that it includes the racemization and *trans* → *cis* isomerization steps in addition to the rearrangement that is of principal interest.

$$\mathbf{K} = \begin{bmatrix} \Sigma\, k & -k_r & 0 & 0 & 0 & 0 \\ -k_r & \Sigma k & 0 & 0 & 0 & 0 \\ -2k_{sr} & -2k_{ai} & 0 & 0 & 0 & 0 \\ -2k_{ar} & -2k_{si} & 0 & 0 & 0 & 0 \\ -2k_{si} & -2k_{ar} & 0 & 0 & 0 & 0 \\ -2k_{ai} & -2k_{sr} & 0 & 0 & 0 & 0 \\ -k_{tc} & -k_{tc} & 0 & 0 & 0 & 0 \end{bmatrix}$$

$$\Sigma\, k = 2(k_{sr} + k_{ar} + k_{si} + k_{ai}) + k_r + k_{tc}$$

The matrix simplifies considerably upon symmetrization:

$$\mathbf{K_s} = \left[\begin{array}{cc|c} \Sigma\, k & -k_r & \\ -k_r & \Sigma\, k & \mathbf{0} \\ \hline & \mathbf{0} & \mathbf{0} \end{array}\right]$$

We can now focus on the nonzero 2×2 submatrix although note that it no longer describes a closed system and so $\lambda_1 \neq 0$.

The submatrix has the same form as that obtained by setting $n = 1$ in the two component generalized solution (Appendix 3) except that the diagonal elements differ by a constant. This constant ($\Sigma\, k - k_r$) will be added to the eigenvalues but its presence will leave the eigenvector matrix unaffected (see Appendix 1). In mechanistic schemes where the various equilibrating species are not destroyed by the same set of irreversible reactions the solution will be somewhat more complex. Under these circumstances it is usually necessary to resort to the determinant method to find the eigenvalues and eigenvectors.

For the present problem the eigenvalues are

$$\lambda_1 = \Sigma\, k - k_r$$

$$\lambda_2 = \Sigma\, k + k_r$$

The eigenvector matrix is

$$\mathbf{B_s} = \frac{1}{\sqrt{2}} \begin{bmatrix} 1 & -1 \\ 1 & 1 \end{bmatrix}$$

The symmetrization matrix is just $\sqrt{2}\,\mathbf{I}$ and so can be ignored. Substitution into Eq. (18) gives the integrated rate equations in the normal way:

$$A_1 = (1/2)(e^{-\lambda_1 t} + e^{-\lambda_2 t}) \tag{23}$$

$$A_2 = (1/2)(e^{-\lambda_1 t} - e^{-\lambda_2 t}) \tag{24}$$

assuming that $A_1^\circ = 1$ and $A_2^\circ = 0$.

These expressions for the time evolution of **1** and **2** can now be substituted into the traditional differential rate expressions for species **3–6**.

$$\frac{dA_3}{dt} = 2k_{sr}A_1 + 2k_{ai}A_2$$

Integrating:

$$A_3 = 2(k_{sr}I_1 + k_{ai}I_2)$$

where $I_1 = \int_0^t A_1\, dt$ and $I_2 = \int_0^t A_2\, dt$.

Similarly,

$$A_4 = 2(k_{ar}I_1 + k_{si}I_2)$$

$$A_5 = 2(k_{si}I_1 + k_{ar}I_2)$$

$$A_6 = 2(k_{ai}I_1 + k_{sr}I_2)$$

These expressions can be rearranged to give the mechanistic rate constants:

$$k_{sr} = \frac{1}{2}\frac{A_3 I_1 - A_6 I_2}{I_1^2 - I_2^2} \tag{25}$$

$$k_{ar} = \frac{1}{2}\frac{A_4 I_1 - A_5 I_2}{I_1^2 - I_2^2} \tag{26}$$

$$k_{si} = \frac{1}{2}\frac{A_5 I_1 - A_4 I_2}{I_1^2 - I_2^2} \tag{27}$$

$$k_{ai} = \frac{1}{2}\frac{A_6 I_1 - A_3 I_2}{I_1^2 - I_2^2} \tag{28}$$

The integrals can be evaluated as follows:

$$I_1 = \int_0^t A_1\, dt$$

$$= \frac{1}{2}\int_0^t (e^{-\lambda_1 t} + e^{-\lambda_2 t})$$

Hence,

$$I_1 = (1/2) \{ \lambda_1^{-1} (1 - e^{-\lambda_1 t}) + \lambda_2^{-1} (1 - e^{-\lambda_2 t}) \} \tag{29}$$

$$I_2 = (1/2) \{ \lambda_1^{-1} (1 - e^{-\lambda_1 t}) - \lambda_2^{-1} (1 - e^{-\lambda_2 t}) \} \tag{30}$$

From Eqs. (23) and (24) one can easily see that the appropriate plot for the determination of λ_1 is $\ln(A_1 + A_2)$ vs. t. This corresponds to following the total rate of disappearance of racemic *trans*-1,2-dipropenylcyclobutane. In order to find λ_2 one plots $\ln(A_1 - A_2)$ vs. t, which corresponds to finding the rate at which *reisolated trans*-1,2-dipropenylcyclobutane has undergone racemization.

These experiments were conducted by Berson's group with the results:

$$\lambda_1 = 1.576 \times 10^{-5} \text{ sec}^{-1}$$

at 146.5°C

$$\lambda_2 = 2.033 \times 10^{-5} \text{ sec}^{-1}$$

Next the reaction mixture was pyrolyzed to completion ($t = $ "∞") and the concentrations of the products **3—6** determined. These results were

$$A_3 = 0.2214$$

$$A_4 = 0.0650$$

$$A_5 = 0.2734$$

$$A_6 = 0.0412$$

Substituting $t = \infty$ into Eqs. (29) and (30) one obtains

$$I_1 = \frac{1}{2} \left\{ \frac{1}{\lambda_1} + \frac{1}{\lambda_2} \right\} = 5.644 \times 10^4 \text{ sec}$$

$$I_2 = \frac{1}{2} \left\{ \frac{1}{\lambda_1} - \frac{1}{\lambda_2} \right\} = 0.701 \times 10^4 \text{ sec}$$

Finally, substituting into Eqs. (25)—(28) one obtains

$$k_{sr} = 1.95 \times 10^{-6} \text{ sec}^{-1} \qquad (41.1\%)$$

$$k_{ar} = 2.80 \times 10^{-7} \text{ sec}^{-1} \qquad (5.9\%)$$

$$k_{ai} = 1.23 \times 10^{-7} \text{ sec}^{-1} \qquad (2.6\%)$$

$$k_{si} = 2.39 \times 10^{-6} \text{ sec}^{-1} \qquad (50.4\%)$$

The figures in parentheses are the proportional contributions of each stereochemical mode to the total rearrangement.

Had the reaction proceeded exclusively via a rotationally equilibrated (achiral) biradical one would have found $k_{sr} = k_{ai}$ and $k_{si} = k_{ar}$. This is clearly not the case. If the 2.6% apparent antarafacial-inversion reaction were in fact the product of an achiral biradical, then so would be 2.6% of the product with apparent suprafacial-retention stereochemistry. But this would mean that the remaining 38.5% of the product with this stereochemistry was formed in some other way, presumably by a forbidden pericyclic reaction. By this interpretation, then, the forbidden concerted pathway is preferred to the biradical pathway by at least 14.8:1.

As always, one could choose to interpret the data in terms of a biradical that did not have time to reach rotational equilibrium. Berson has advanced arguments against this interpretation, however.[18]

The kinetic problem of two interconverting enantiomers proceeding irreversibly to a number of products has been treated by a different mathematical technique[19] but the approach outlined here is more general in that it is not restricted to enantiomers and can, in principle, handle any number of equilibrating precursors.

4.6. NUMERICAL INTEGRATION OF RATE EQUATIONS

The mathematical technique described in Section 4.5 is a powerful tool for handling complex kinetic schemes but it obviously has its limitations. Nonunimolecular reactions are not easily dealt with by this approach, for example, and, even within the class of unimolecular networks, reactions involving reversible interconversion of several chemically different compounds (i.e., not label isomers or enantiomers) often present a problem. When faced with these difficulties one is usually forced to resort to numerical integration techniques. We will discuss two main types. The first is based on the matrix approach discussed in Section 4.5. The second involves direct integration of the differential rate equations. The advantages and limitations of the two methods will be discussed.

4.6.1. Matrix Methods

If one tries to apply the techniques described in Section 4.5 to a problem involving reversible interconversion of several species that differ in their enthalpy of formation one usually finds that it is not possible to determine the eigenvalues and eigenvectors of the rate constant matrix. There are exceptions. For example, the generalized three component solution listed in Appendix 3 is still valid even when compound **2** is chemically different from **1** and **3**. One merely replaces the parameter n by the experimentally determined equilibrium constant. Unfortunately these cases are relatively rare. For the less cooperative ones numerical techniques can be helpful.

The need to find the eigenvalues and eigenvectors of a real symmetric matrix is commonly encountered in chemistry. It is central to most molecular orbital calculations, for example. Consequently there are a number of computer algorithms for handling the problem, including the Jacobi rotation[20] and Householder-Givens[21] techniques. The major limitation, for our purposes, is that these algorithms can only be applied to matrices whose elements are numerical constants—the computer cannot find the eigenvalues and eigenvectors of a matrix whose elements are algebraic variables. One is thus forced to follow the rather tedious technique of guessing at values for the mechanistic rate constants and then using the program to calculate the resulting time evolutions of the observable species. The values of the mechanistic rate constants are then adjusted (manually or automatically) until the calculated time evolutions are within some acceptable tolerance of the experimentally determined ones. Symmetry considerations and the requirements of microscopic reversibility (see Appendix 2) place some limitations on the allowable values for the mechanistic rate constants but the tedium involved in following this approach can still be extreme.

A somewhat superior method is to make use of the fact that a unimolecular network of n interconverting species always has integrated rate equations of the form:

$$A_i = \sum_{j=1}^{n} c_{ij} e^{-\lambda_j t} \qquad i = 1, 2, \cdots n$$

regardless of the chemical identity of the species involved. Accordingly, one can use nonlinear least-squares methods[22] to find the best set of c_{ij} and λ_j for a given set of experimental data. It is then possible to work backward through the algebra described in Section 4.5 to deduce a unique set of mechanistic rate constants.

4.6.2. Direct Numerical Integration of Rate Equations

Direct numerical integration techniques again suffer from the need to make repeated guesses at the values of mechanistic rate constants. They have the additional disadvantage that computation times are usually much longer than those required for the matrix method and the results are sometimes less accurate, as we will see in the following discussion. Their big advantage is generality— direct numerical integration can be applied to differential rate equations of any order with no significant difference in the method or difficulty of the computation. The general strategy and the reasons for some of the problems can best be illustrated by beginning with a discussion of the very simplest numerical integration technique.

Consider the following reaction scheme:

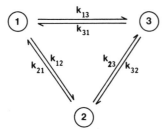

The rate laws for disappearance of 1 and 2 are

$$\frac{-dA_1}{dt} = (k_{12} + k_{13})A_1 - \qquad k_{21}A_2 - k_{31}A_3$$

$$\frac{-dA_2}{dt} = \qquad - k_{21}A_1 + (k_{21} + k_{23})A_2 - k_{32}A_3$$

These differential equations can be written as approximate finite difference equations for some small time interval Δt:

$$-\Delta A_1 = \{(k_{12} + k_{13})A_1 - \qquad k_{21}A_2 - k_{31}A_3\}\Delta t \qquad (31)$$

$$-\Delta A_2 = \{ \qquad - k_{21}A_1 + (k_{21} + k_{23})A_2 - k_{32}A_3\}\Delta t \qquad (32)$$

The accuracy of Eqs. (31) and (32) depends on the size of the time interval Δt. Unfortunately, as we shall see in the following example, the computation time also depends on the size of Δt. The accuracy of the final answer and the time required to achieve it both increase as Δt decreases. Consider the example where $A_1^{\circ} = 1.0$; $A_2^{\circ} = A_3^{\circ} = 0.0$ and the rate constants have the following values:

$$k_{12} = 1.0 \times 10^{-5} \text{ sec}^{-1} \qquad k_{21} = 0.5 \times 10^{-5} \text{ sec}^{-1}$$

$$k_{23} = 1.5 \times 10^{-5} \text{ sec}^{-1} \qquad k_{13} = 3.0 \times 10^{-5} \text{ sec}^{-1}$$

$$k_{31} = 2.0 \times 10^{-5} \text{ sec}^{-1} \qquad k_{32} = 2.0 \times 10^{-5} \text{ sec}^{-1}$$

Let us assume that we wish to know the concentrations of **1, 2,** and **3** at $t = 10,000$ sec. We will use a step size, Δt, of 500 sec. Substituting into Eqs. (31) and (32):

$$\Delta A_1 = -\{(1.0 + 3.0) \times 1.0 - 0.5 \times 0.0 - 2.0 \times 0.0\}$$

$$\times 10^{-5} \times 500$$

$$= -0.0200$$

$$\Delta A_2 = -\{-1.0 \times 1.0 + (0.5 + 1.5) \times 0.0 - 2.0 \times 0.0\}$$

$$\times 10^{-5} + 500$$

$$= +0.0050$$

$$A_1(500) = A_1^\circ + \Delta A_1$$

$$= 0.9800$$

$$A_2(500) = A_2^\circ + \Delta A_2$$

$$= 0.0050$$

$$A_3(500) = A_1^\circ + A_2^\circ + A_3^\circ - A_1(500) - A_2(500)$$

$$= 0.0150$$

The next step is taken by re-evaluating Eqs. (31) and (32) using $A_i(500)$ in place of A_i°. The results are

$$\Delta A_1 = -0.0914$$

$$\Delta A_2 = 0.0050$$

Hence

$$A_1(1000) = 0.9606$$

$$A_2(1000) = 0.0100$$

$$A_3(1000) = 0.0294$$

Evaluation of $A_i(10,000)$ requires 18 more such calculations. The results are

$$A_1(10,000) = 0.6900$$

$$A_2(10,000) = 0.0974$$

$$A_3(10,000) = 0.2127$$

The exact solutions can be obtained by using the matrix method. They are (to four decimal places)

$$A_1(10,000) = 0.6931$$

$$A_2(10,000) = 0.0971$$

$$A_3(10,000) = 0.2098$$

The error in the direct numerical integration arises because A_1, A_2, and A_3 are assumed to be constant throughout the time interval Δt. The situation is depicted in Figure 4.9. During the first interval Δt the concentration of **1** is set at 1.0 while those of **2** and **3** are set at 0.0. In reality, of course, A_1 decreases during this time while A_2 and A_3 increase. The upshot is that our approximation leads to an overestimate of the rate at which compound **1** gives **2** and **3**, with the result that A_1 is too low while A_2 and A_3 are too high at the end of the time interval.

One could hope to improve this situation by using a slightly smaller constant value for A_1 and slightly larger values for A_2 and A_3 during the first time interval. Reasonable values might be the average of A_i^0 and $A_i(\Delta t)$. In other words, one would make two estimates of the concentrations during each time interval:

$$\Delta A_{1j} = -\{(k_{12} + k_{13})A_{1j} - k_{21}A_{2j} - k_{31}A_{3j}\}\Delta t$$

$$\Delta A_{2j} = -\{-k_{21}A_{1j} + (k_{21} + k_{23})A_{2j} - k_{32}A_{3j}\}\Delta t$$

$$A_{1j}(\text{av}) = \{(A_{1j-1} + \Delta A_{1j}) + A_{1j-1}\}/2$$

$$= A_{1j-1} + \Delta A_{1j}/2$$

$$A_{2j}(\text{av}) = A_{2j-1} + \Delta A_{2j}/2$$

$$\Delta' A_{1j} = -\{(k_{12} + k_{13})A_{1j}(\text{av}) - k_{21}A_{2j}(\text{av}) - k_{31}A_{3j}(\text{av})\}\Delta t$$

$$A_{1j} = A_{1j-1} + \Delta' A_{1j}$$

$$A_{2j} = A_{2j-1} + \Delta' A_{2j}$$

ΔA_{ij} is the first estimate of the change in concentration of species i during the jth time interval, $\Delta' A_{ij}$ is the second estimate of that change, and A_{ij} is the concentration of species i after the jth interval. $A_{ij}(\text{av})$ is the fixed value that approximates the concentration of species i during the jth time interval.

Using these equations with the numbers of the previous example one finds:

$$A_1(500) = 0.9803$$

$$A_2(500) = 0.0050$$

$$A_3(500) = 0.0147$$

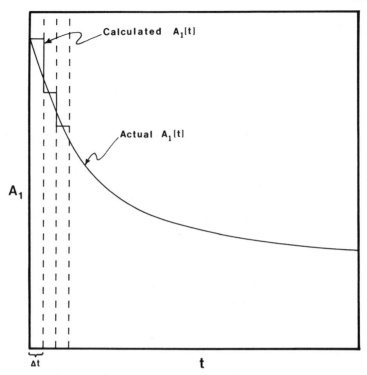

FIGURE 4.9. True time evolution of A_1 and the approximation to it used in the simplest form of numerical integration.

and, after 19 more sets of calculations:

$$A_1(10{,}000) = 0.6931$$

$$A_2(10{,}000) = 0.0971$$

$$A_3(10{,}000) = 0.2098$$

One might wonder whether the extra complexity of the calculations required for this second integration method defeat the original purpose. Would it not have been just as good to use the first method with a smaller step size? The answer turns out to be "no." In order to attain results accurate to the fourth decimal place for the present problem the first method requires about 30 times longer than the second on a programmable calculator.

Clearly one can go on improving the estimates of $A_{ij}(\text{av})$ by making more and more calculations of ΔA_{ij} for each time interval. The trade-off in complexity of calculation for each step vs. number of steps required for a given degree of accuracy seems to be optimized in the so-called Runge-Kutta methods.[23] The most commonly used of these for problems in reaction kinetics is the fourth-order Runge–Kutta integration which, as its name implies, makes four estimates of ΔA_{ij}. For the present problem the equations would be:

$$\Delta_1 A_{1j} = -\{(k_{12} + k_{13})A_{1j} - k_{21}A_{2j} - k_{31}A_{3j}\}\Delta t$$

$$\Delta_2 A_{1j} = -\{(k_{12} + k_{13})(A_{1j} + \Delta_1 A_{1j}/2) - k_{21}(A_{2j} + \Delta_1 A_{2j}/2)$$

$$-k_{31}(A_{3j} + \Delta_1 A_{3j}/2)\}\Delta t$$

$$\Delta_3 A_{1j} = -\{(k_{12} + k_{13})(A_{1j} + \Delta_2 A_{1j}/2) - k_{21}(A_{2j} + \Delta_2 A_{2j}/2)$$

$$-k_{31}(A_{3j} + \Delta_2 A_{3j}/2)\}\Delta t$$

$$\Delta_4 A_{1j} = -\{(k_{12} + k_{13})(A_{1j} + \Delta_3 A_{1j}) - k_{21}(A_{2j} + \Delta_3 A_{2j})$$

$$-k_{31}(A_{3j} + \Delta_3 A_{3j})\}\Delta t$$

$$A_{1j} = A_{1j-1} + (\Delta_1 A_{1j} + 2\Delta_2 A_{1j} + 2\Delta_3 A_{1j} + \Delta_4 A_{1j})/6$$

REFERENCES

1. A. A. Frost and R. G. Pearson, *Kinetics and Mechanism*, 2nd ed., Wiley, New York, 1961; I. Amdur and G. G. Hammes, *Chemical Kinetics: Principles and Selected Topics*, McGraw-Hill, New York, 1966; J. H. Espenson, *Chemical Kinetics and Reaction Mechanisms*, McGraw-Hill, New York, 1981.

2. E. D. Hughes, C. K. Ingold, and C. S. Patel, *J. Chem. Soc.*, 526 (1933).

3. L. C. Bateman, E. D. Hughes, and C. K. Ingold, *J. Chem. Soc.*, 974 (1940).

4. W. J. Moore, *Physical Chemistry*, 3rd ed., Prentice-Hall, Englewood Cliffs, N.J., 1962, p. 368.

5. G. S. Hammond, *J. Am. Chem. Soc.*, **77**, 334 (1955).

6. W. G. Young, S. Winstein, and H. L. Goering, *J. Am. Chem. Soc.*, **73**, 1958 (1951).

7. S. Winstein and G. C. Robinson, *J. Am. Chem. Soc.*, **80**, 169 (1958).

8. S. Winstein, P. E. Kleindinst, Jr., and G. C. Robinson, *J. Am. Chem. Soc.*, **83**, 885 (1961).

9. F. A. Matsen and J. L. Franklin, *J. Am. Chem. Soc.*, **72**, 3337 (1950).

10. C. D. Ritchie, *Physical Organic Chemistry, the Fundamental Concepts*, Dekker, New York, 1975, pp. 7–27.

11. M. J. Goldstein, M. S. Benzon, W. A. Haiby, and H. A. Judson, in *Mechanisms of Hydrocarbon Reactions*, F. Marta and D. Kallo (Eds.), Akademiai Kiado, Budapest, 1975, pp. 779–810.

12. F. T. Smith, *J. Chem. Phys.*, **29**, 235 (1958).

13. R. Hoffmann, *J. Am. Chem. Soc.*, **90**, 1475 (1968).

14. S. W. Benson, *J. Chem. Phys.*, **34**, 521 (1961); H. E. O'Neal and S. W. Benson, *J. Phys. Chem.*, **72**, 1866 (1968).

15. J. A. Berson, L. D. Pedersen, and B. K. Carpenter, *J. Am. Chem. Soc.*, **98**, 122 (1976).

16. M. R. Willcott, III and V. H. Cargle, *J. Am. Chem. Soc.*, **91**, 4310 (1969).

17. J. E. Baldwin and C. G. Carter, *J. Am. Chem. Soc.*, **101**, 1325 (1979).

18. J. A. Berson, P. B. Dervan, R. Malherbe, and J. A. Jenkins, *J. Am. Chem. Soc.*, **98**, 5937 (1976).

19. S. Winstein and D. Trifan, *J. Am. Chem. Soc.*, **74,** 1154 (1952).

20. K. B. Wiberg, *Computer Programming for Chemists*, Benjamin, New York, 1965, p. 48.

21. J. W. Givens, *J. Assoc. Comput. Mach.*, **4,** 298 (1957).

22. W. E. Wentworth, *J. Chem. Educ.*, **42,** 96 (1965).

23. W. E. Milne, *Numerical Solution of Differential Equations*, Wiley, New York, 1953, pp. 72–73.

CHAPTER 5

ISOTOPE EFFECTS

In the preceding chapters we have tended to treat all isotopes of a given element as chemically identical. For many purposes this is a good approximation, but there are circumstances under which different isotopes do show different chemical reactivity. This chapter will contain a brief discussion of the causes of such isotope effects and a more detailed discussion of the mechanistic information that can be acquired from their measurement.

Isotope effects can conveniently be divided into two classes: equilibrium isotope effects such as

$$HT + H_2O \rightleftharpoons H_2 + HTO \qquad K_{298} = 6.26 \pm 0.10$$

$$H^{12}CN + {}^{13}CN^- \rightleftharpoons H^{13}CN + {}^{12}CN^- \qquad K_{298} = 1.026 \pm 0.002$$

and kinetic isotope effects such as

We will spend most time discussing the kinetic type but development of the theory will begin with an analysis of the equilibrium type.

5.1. THEORY OF ISOTOPE EFFECTS

According to the Born–Oppenheimer approximation the electronic and nuclear wavefunctions for an atom or molecule can be treated as being independent. Consequently isotopic substitution should not affect the potential energy surface for a reaction. Differences in reactivity of isotopically related molecules must, therefore, be due to differences in the vibrational and rotational energy levels that lie above the potential energy surface. In order to determine what these differences are we will appeal to some elementary statistical mechanics.

The equilibrium constant for the interconversion of two molecules A and B can be written

$$K = \frac{Q_B e^{-[\epsilon_0(B) - \epsilon_0(A)]/kT}}{Q_A}$$

$$Q = \sum_i g_i e^{-\epsilon_i/(kT)}$$

where Q = partition function
       ϵ_0 = lowest energy level
       k = Boltzmann's constant
       g_i = statistical weight (degeneracy) of energy level i

Again applying the Born–Oppenheimer approximation we can divide the total partition function into translational, rotational, vibrational, electronic, and nuclear contributions:

$$Q_{tot} = Q_{trans} Q_{rot} Q_{vib} Q_{elec} Q_{nuc}$$

Treating the electronic part first:

$$Q_{elec} = e^{-\epsilon_0/(kT)} + e^{-\epsilon_1/(kT)} + \cdots$$

If we define $\epsilon_0 = 0$ then, for most organic molecules at typical reaction temperatures, $\epsilon_1 \gg kT$. Hence

$$Q_{elec} \approx 1$$

provided that the ground electronic state is singly degenerate (as is usually the case). The contribution of the nuclear partition function can be neglected for the same reason. The total partition function can thus be considered to be a product of the translational, rotational, and vibrational contributions. The quantum mechanical expressions for these partition functions can be deduced from solution of the Schrödinger equation for various simplified models.[1]

$$Q_{\text{trans}} = \frac{(2\pi M \mathbf{k} T)^{3/2} V}{h^3}$$

where M = mass of molecule
V = volume of container
h = Planck's constant

The expression for the rotational partition function comes from the rigid rotor approximation for a nonlinear molecule

$$Q_{\text{rot}} = \frac{\{\pi(8\pi^2 \mathbf{k} T)^3 I_A I_B I_C\}^{1/2}}{\sigma h^3}$$

where I_A = moment of inertia about A axis
σ = symmetry number

The approximation for the vibrational partition function comes from the solutions to the Schrödinger equation for a harmonic oscillator

$$Q_{\text{vib}} = \prod_i^{3n-6} \left\{ \frac{e^{-h\nu_i/(2\mathbf{k} T)}}{(1 - e^{-h\nu_i/(\mathbf{k} T)})} \right\}$$

where ν_i = frequency of normal mode i
n = number of atoms in molecule

The parameters M, I, and ν in these expressions are all mass dependent and hence sensitive to isotopic substitution. It is the mass dependence of the partition functions that is responsible for both equilibrium and kinetic isotope effects.

For small molecules whose vibrational frequencies and moments of inertia are known, the preceding equations allow quite accurate calculation of equilibrium isotope effects. The equilibrium constants calculated for the HT + H_2O and $H^{12}CN$ + $^{13}CN^-$ reactions mentioned in the introduction are 6.35 and 1.030, for example.

Extension of this approach to the calculation of kinetic isotope effects is most easily achieved by use of transition state theory.[2] The two assumptions that are central to this model are that the reactant is at equilibrium with the activated complex (transition state) and that one of the vibrational modes of the reactant becomes a kind of translation (passage across the top of the potential energy barrier) in the transition state. We can write equations for the rate constants for reaction of the two isotopically related molecules:

$$k_1 = \kappa_1 \, \frac{kT}{h} \frac{Q_1^{\ddagger}}{Q_1} \, e^{(\epsilon_0 - \epsilon_0^{\ddagger})/(kT)}$$

$$k_2 = \kappa_2 \, \frac{kT}{h} \frac{Q_2^{\ddagger}}{Q_2} \, e^{(\epsilon_0 - \epsilon_0^{\ddagger})/(kT)}$$

where k_1 = rate constant for reaction of isotopically lighter molecule
        k_2 = rate constant for reaction of isotopically heavier molecule
        κ = transmission coefficient

The exponents are identical for the two equations, in keeping with the belief that isotopic substitution has no effect on the potential energy surface. Expanding the partition functions as before, the ratio k_1/k_2 can be written:

$$\frac{k_1}{k_2} = \frac{\kappa_1}{\kappa_2} \frac{\sigma_1}{\sigma_2} \frac{\sigma_2^{\ddagger}}{\sigma_1^{\ddagger}} \cdot \text{MMI} \cdot \text{EXC} \cdot \text{ZPE} \tag{1}$$

where   MMI = mass, moment of inertia term

$$= \frac{(M_1^{\ddagger} M_2)^{3/2} (I_{A1}^{\ddagger} I_{B1}^{\ddagger} I_{C1}^{\ddagger} I_{A2} I_{B2} I_{C2})^{1/2}}{(M_2^{\ddagger} M_1)^{3/2} (I_{A2}^{\ddagger} I_{B2}^{\ddagger} I_{C2}^{\ddagger} I_{A1} I_{B1} I_{C1})^{1/2}}$$

EXC = excitation term

$$= \frac{\displaystyle\prod^{3n-6} \left\{ (1 - e^{-u_{i1}})/(1 - e^{-u_{i2}}) \right\}}{\displaystyle\prod^{3n-7} \left\{ (1 - e^{-u_{i1}^{\ddagger}})/(1 - e^{-u_{i2}^{\ddagger}}) \right\}}$$

ZPE = zero point energy term

$$= \frac{\exp\left\{ \displaystyle\sum^{3n-6} (u_{i1} - u_{i2})/2 \right\}}{\exp\left\{ \displaystyle\sum^{3n-7} (u_{i1}^{\ddagger} - u_{i2}^{\ddagger})/2 \right\}}$$

$$u_i = \frac{h\nu_i}{kT}$$

Note the limit of $3n - 7$ on the summations and products over transition state vibrations. The "missing" vibration is the one that has become the reaction coordinate.

Equation (1) can be simplified considerably by making some approximations. If we assume that $\kappa_1 \simeq \kappa_2$, $M_2^{\ddagger}/M_1^{\ddagger} \simeq M_2/M_1$, and $I_2^{\ddagger}/I_1^{\ddagger} \simeq I_2/I_1$ then

MMI $\simeq 1$. Furthermore since most organic molecules have vibration frequencies $> 500\,\mathrm{cm}^{-1}$, $e^{-u_i} < 0.08$ at $25°\mathrm{C}$ thus EXC $\simeq 1$. Hence the dominant contributor to the kinetic isotope effect is seen to be the zero point energy term:

$$\frac{k_1}{k_2} \simeq \frac{\sigma_1 \sigma_2^{\ddagger} \exp\left\{\sum^{3n-6} (u_{i1} - u_{i2})/2\right\}}{\sigma_2 \sigma_1^{\ddagger} \exp\left\{\sum^{3n-7} (u_{i1}^{\ddagger} - u_{i2}^{\ddagger})/2\right\}} \qquad (2)$$

Remembering the equations from solution of the Schrödinger equation for a diatomic harmonic oscillator:

$$\epsilon_0 = h\nu/2$$

$$\nu = \frac{1}{2\pi}\sqrt{(f/\mu)}$$

$$f = \text{force constant}$$

$$\mu = \text{reduced mass}$$

$$= m_1 m_2 / (m_1 + m_2)$$

we can make the following statements:

1. The isotopically heavier molecule always has the lower zero point energy.
2. The difference in zero point energy between the two isotopically related molecules increases as the force constant of the bond increases.
3. As a consequence of statements 1 and 2, the heavier isotope will prefer to be in the more strongly bound location in a polyatomic molecule.
4. The difference in zero point energy is largest for the hydrogen isotopes and decreases with increasing atomic weight. Kinetic isotope effects will show the same behavior.

5.2. PRIMARY KINETIC ISOTOPE EFFECTS

Primary kinetic isotope effects occur in reactions that involve cleavage of a bond to an isotopically labeled atom. We will spend most time discussing H/D isotope effects since these are the most frequently encountered, but towards the end of this section we will also consider the information that can be gained by study of heavy atom isotope effects.

The simplest model for a primary H/D isotope effect would be one in which the C—H or C—D bond is treated as an isolated diatomic molecule. In the case where all symmetry numbers are unity, this model leads to a considerable simplification of Eq. (2):

$$\frac{k_H}{k_D} = \frac{k_1}{k_2} \simeq \exp\left\{\frac{hc(\tilde{\nu}_1 - \tilde{\nu}_2)}{2kT}\right\}$$

For a diatomic molecule there is only one vibration and in the transition state there are none, the latter giving the denominator of Eq. (2) a value of unity. Typical values for $\tilde{\nu}_1$ and $\tilde{\nu}_2$ are 2900 and 2100 cm^{-1}, respectively, leading to a value of 6.9 for k_H/k_D at 25° C. Of course this model does not necessarily imply that one will observe $k_H/k_D \simeq 6.9$ in all reactions that involve cleavage of a C—H bond. *A primary isotope effect can be detected by kinetic measurement only if the bond to the isotopically labeled atom is broken during or before the rate-determining step.* Consider, for example, the electrophilic substitution of benzene:

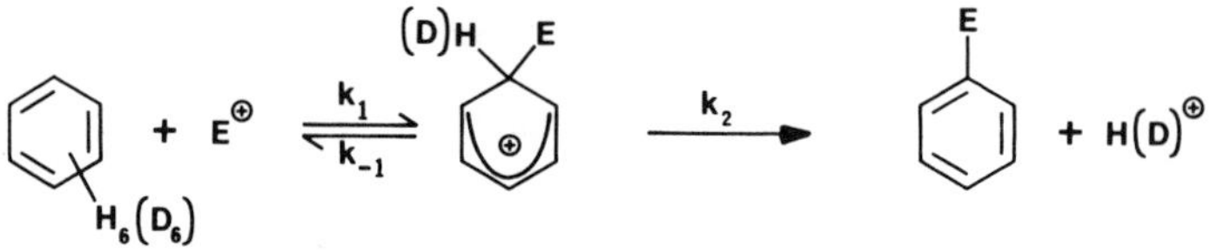

The form of the differential rate equation can be deduced by applying the steady-state approximation to the cyclohexadienyl cation intermediate:

$$\frac{-d[\text{benzene}]}{dt} = \frac{k_1 k_2 [\text{benzene}][\text{E}^+]}{k_{-1} + k_2}$$

For reactions in which $k_2 \gg k_{-1}$ (i.e., k_1 is rate determining)

$$\frac{-d[\text{benzene}]}{dt} = k_1 [\text{benzene}][\text{E}^+]$$

Now the phenomenological rate constant k_{obs} can be equated with the mechanistic rate constant k_1. The rate-determining step in this case does not involve cleavage of a C—H bond and so k_{obs} should have about the same value when C_6D_6 is used in place of C_6H_6, that is, $k_H/k_D \simeq 1$. This is the experimental observation when $\text{E}^+ = \text{NO}_2^+$.

For reactions in which $k_{-1} \gg k_2$ (i.e., k_2 is rate determining) the rate equation becomes

$$\frac{-d[\text{benzene}]}{dt} = \frac{k_1}{k_{-1}} \cdot k_2 [\text{benzene}][\text{E}^+]$$

Now the phenomenological rate constant can be identified with the product Kk_2 where $K = k_1/k_{-1}$. In this situation one would expect a primary isotope effect on

k_2 to appear in k_{obs}. From the model discussed above we would predict $k_H/k_D \approx 6.9$ at 25°C. For the case where $E^+ = Hg^{2+}$ the experimental value is 6.75.

Thus one important use of isotope effects is to determine which step in a complex mechanism is rate determining. Unfortunately the model that we have used to predict primary isotope effects makes this task look simpler than it really is.

From our discussion so far one would expect to see $k_H/k_D \approx 1$ for reactions in which C—H cleavage occurs after the rate determining step and $k_H/k_D \approx 6.9$ for reactions in which C—H cleavage occurs during or before the rate determining step. The model does not allow for any other possibilities. The experimental facts are that one finds values of k_H/k_D ranging anywhere between 1 and 35 at 25°C. Clearly some improvements in the model are required.

One indication of an important deficiency in the model comes from a study of halogen atom reactions with toluene:[3]

When the methyl hydrogens are replaced by deuterium one finds $k_H/k_D = 1.5$ for X = Cl but $k_H/k_D = 4.6$ for X = Br. It is apparent that the halogen atom is involved in the transition state for the reaction and should be included in the model for the isotope effect. Indeed, on reflection, one would probably be hard pressed to think of many reactions that involve true unimolecular cleavage of a C—H bond as implied in the first model. Even the electrophilic substitution of benzene almost certainly involves transfer of a hydrogen ion from the intermediate to a solvent molecule, not loss of a "naked" proton. Thus a minimal realistic model for a hydrogen/deuterium primary isotope effect must involve a three-body transition state. The implications of such a model were first discussed in detail by Westheimer[4] and his arguments are presented (in considerably abbreviated form) below.

We will first consider the situation in which a hydrogen (it does not matter whether it is a hydrogen atom or a proton) is transferred from the reactant to some reagent, B, by way of a linear transition state:

$$R_3C—H + :B \longrightarrow R_3C---H---B^{\ddagger} \longrightarrow R_3C: + H—B$$

If we restrict our attention to the vibrations of the three atom assembly C—H—B then we can identify four normal modes, two stretching and two (degenerate) bending vibrations. Concentrating first on the stretching motions, we can further recognize that the antisymmetric mode $r_1 - r_2$ is really the reaction coordinate that transfers the hydrogen from C to B and back again. In accord with the precepts of transition state theory this motion must be assumed to have an imaginary frequency and a negative restoring force.

$$r_1 \qquad r_2$$

Antisymmetric: $C \longleftrightarrow H \rightarrowtail\!\!\!\dashv B$ $r_1 - r_2$

Symmetric: $C \longleftrightarrow H \longleftrightarrow B$ $r_1 + r_2$

The symmetric mode $r_1 + r_2$ is a real vibration that can be represented as an orthogonal coordinate on a three-dimensional potential energy diagram (see Figure 5.1). The characteristic frequency for the symmetric mode, $\nu_s^{\ddagger}$, can be deduced from classical mechanics if one treats the two bonds as coupled harmonic oscillators. The result is

$$\nu_s^{\ddagger} = \frac{1}{2\pi} \left\{ \frac{f_1}{m_C} + \frac{f_2}{m_B} + \frac{f_1 + f_2 - 2\sqrt{(f_1 f_2)}}{m_H} \right\}^{1/2} \tag{3}$$

This frequency will appear in the exponent of the denominator of Eq. (2) and will tend to reduce the magnitude of k_H/k_D. Inclusion of the bending vibrations would further reduce the isotope effect. How, then, can one observe H/D isotope effects approaching the value of 6.9 deduced from the first model? The

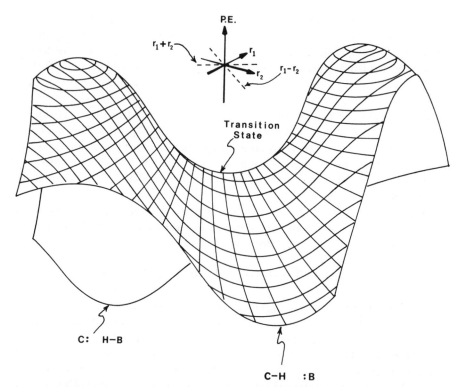

FIGURE 5.1. Potential energy diagram for the reaction $C—H + B \longrightarrow C + B—H$.

answer becomes apparent upon closer inspection of Eq. (3). Under conditions where $f_1 = f_2$, the third term in the summation vanishes. The frequency then becomes *insensitive to the mass of the hydrogen atom*, that is, insensitive to replacement of H by D. This means that the exponent $u_s^{\ddagger}(H) - u_s^{\ddagger}(D)$ in the denominator of Eq. (2) goes to zero. We thus deduce that the isotope effect for a linear hydrogen transfer should be at a maximum when the transition state is symmetrically placed between reactants and products—in other words when the hydrogen is equally bonded to the reactant and the reagent. Very exothermic reactions (in which the transition state tends to occur early in the reaction coordinate) would be expected to have smaller values of k_H/k_D. Empirical observation supports this prediction.[4] It is presumably this kind of trend that is responsible for the difference in isotope effect between the reactions of Br· and Cl· with toluene.

When the transition state for H transfer becomes nonlinear the bending and stretching vibrations of the three atom assembly are no longer separable by symmetry. The symmetric stretching mode that used to be relatively insensitive to isotopic substitution now becomes mixed with a bending mode in which the hydrogen executes a large amplitude vibration. The sensitivity of this mode to change in mass of the hydrogen isotope increases, and the denominator of Eq. (2) increases accordingly. The result is a reduction in the observed value of k_H/k_D.

The three-body model for primary H/D isotope effects is thus a distinct improvement over the original two-body model in that we can now see how the observed value of k_H/k_D will depend on the geometry and energetics of the reaction. But, as one might expect, even the three-body model has its limitations. In particular, the upper limit calculated for k_H/k_D is still 6.9 at 25° C, whereas the experimental upper limit is close to 35.

Part of this discrepancy can be resolved by including the other vibrations of the reactants and transition state in the calculation. For polyatomic molecules this can be a task of some substantial magnitude and so various approximations or "cut-off" methods have been devised whereby vibrations involving atoms that are more than two or three bonds removed from the reaction center can be ignored.[5] These approximations often give results that are close to those from a full calculation but not necessarily so close to the experimental observation. The difficulty, of course, is that one has to guess at a set of vibration frequencies for the transition state since it is not directly observable. This limitation means that kinetic isotope effects can never be predicted with as much accuracy as equilibrium isotope effects can.

Even when all the other vibrations in the molecules are included the upper limit calculated for k_H/k_D rarely exceeds 18 at 25° C. In order to explain values much larger than this it is necessary to consider the phenomenon of quantum mechanical tunneling.

Two aspects of the tunneling phenomenon result in its contributing to the H/D isotope effect. The first is that the vibrational wavefunction has a mass dependent exponent that leads to a more rapid fall off in Ψ^2 as the reduced mass of the oscillator increases.

$$\Psi = (\pi/2\beta)^{1/4} \exp\{-(\pi x^2 \sqrt{\mu f})/h\}$$

$$\beta = 2\pi^2 \mu \nu / h$$

$$x = R - R_{eq}$$

As a consequence the penetration of the barrier is smaller for a C—D vibrational wavefunction than it is for a C—H. The second contributor is that the consistently higher energy for the C—H levels means that the barrier is a little narrower for C—H than it is for C—D at the same vibrational quantum number (see Figure 5.2). Both aspects of the phenomenon lead one to expect a slower tunneling for C—D than for C—H, resulting in an increase of k_H/k_D. This is the observed direction of the effect.[6]

Tunneling contributions to the isotope effect are maximized by low temperature (which decreases the classical contribution since fewer molecules have sufficient energy to surmount the barrier) and narrow barriers (usually meaning a reaction in which the distance traveled by the proton is small). A particularly startling example that incorporates both of these features has been

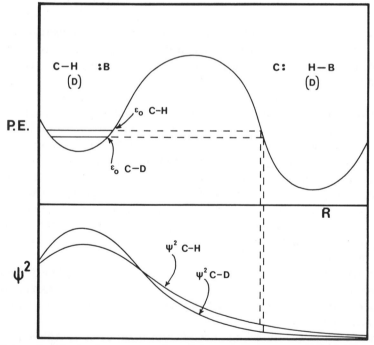

FIGURE 5.2. Contribution from quantum mechanical tunneling to a primary H/D isotope effect. For vibrations of the same quantum number the C—H level is a little higher and the barrier is a little narrower. Furthermore Ψ^2 falls of with distance faster for C—D than for C—H. Both of these effects give C—H a higher probability of being in the product potential well.

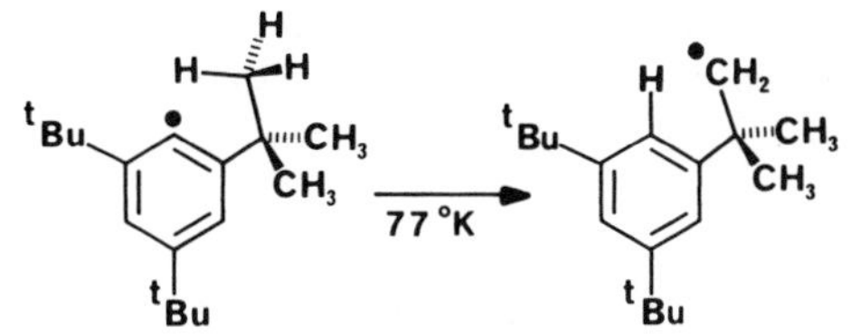

provided by Ingold[7] who found that the reaction shown above has $k_H/k_D \approx 13,000$.

Models that allow quantitative prediction of tunneling contributions to the isotope effect have been reported[8] but they will not be discussed here.

To summarize, primary kinetic isotope effects are useful in determining which step in a complex reaction mechanism is rate determining. Given suitable standards for comparison, they can also provide information about the geometry of the transition state and its position along the reaction coordinate.

The discussion so far has concentrated on H/D isotope effects but useful information can also be gained by measurement of heavy atom isotope effects. The experiments are technically more difficult simply because of the small magnitude of the effects but there are circumstances under which no other technique is appropriate. An example is the mechanistic investigation of the Baeyer-Villiger oxidation conducted by Fry and coworkers.[9]

Acetophenone reacts with m-chloroperbenzoic acid to give phenyl acetate. As shown in Figure 5.3, one could imagine two mechanisms for this process that differ just in the timing of the bond-breaking processes. Both involve the tetrahedral intermediate 1 but mechanism A leads to the formation of an oxenium ion (2) whereas mechanism B involves concerted phenyl migration and loss of

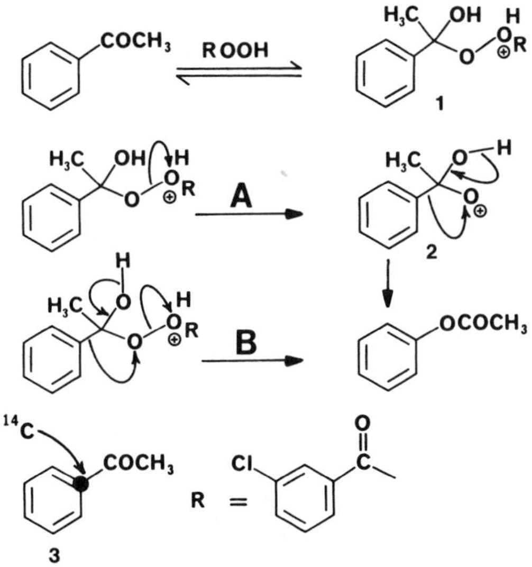

FIGURE 5.3. Mechanisms for the Baeyer–Villiger oxidation of acetophenone.

m-chlorobenzoic acid. Formation of **2** would almost certainly be the rate-determining step in mechanism A, and so one could hope to distinguish between A and B by looking for a heavy atom isotope effect using the *ipso*-labeled reactant **3**.

The experimental observation that $k_{12_C}/k_{14_C} = 1.048 \pm 0.002$ at 32°C provides clear support for mechanism B over mechanism A.

The subtlety of mechanistic information that can be garnered from isotope effect studies is well illustrated by the work of Cram[10] and Hofmann[11] on the formation of carbanions in hydroxylic solvents. The impetus for the study was the observation that **4** could be made to aromatize in a deuterated medium with amounts of deuterium incorporation that varied from 2 to 83% (see Figure 5.4). Furthermore, **5**, when subjected to a racemization/exchange study exhibited $k_\alpha/k_{ex} \approx 10$. Our earlier discussion of racemization/exchange experiments (Section 3.4) led us to expect an upper limit of 1 for k_α/k_{ex}.

Both experimental results could be explained by a mechanism in which the carbanion remains hydrogen bonded to the protonated base and is able to undergo rearrangement in this state:

$$R\!-\!D + B^- \rightleftharpoons \{R\text{----}D\text{---}B\}^- \longrightarrow \{R\text{----}H\text{---}B\}^-$$

$$\downarrow \qquad\qquad\qquad\qquad \downarrow$$

racemize or   exchange
rearrange

An experimental test for this hypothesis is provided by an isotope effect study.

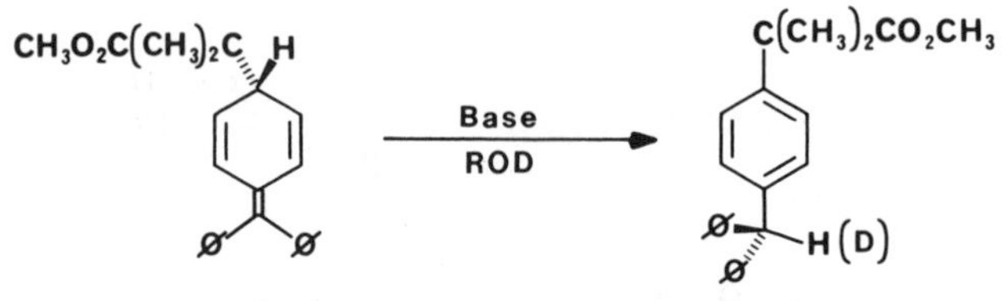

Base	Solvent	% D
DOCH$_2$CH$_2$OK	DOCH$_2$CH$_2$OD	83
CH$_3$ONa	CH$_3$OD	54
i-Pr$_3$N	$(C_2H_5)_3$COD	2

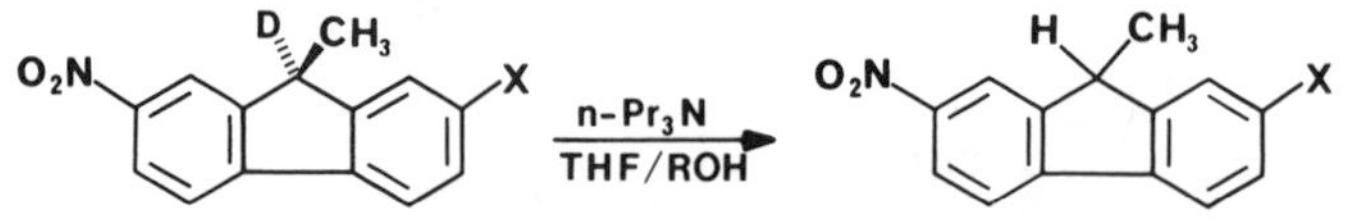

$X = CON(CH_3)_2$

FIGURE 5.4. Experimental results that seem to require the intermediacy of a carbanion/(protonated base) complex.

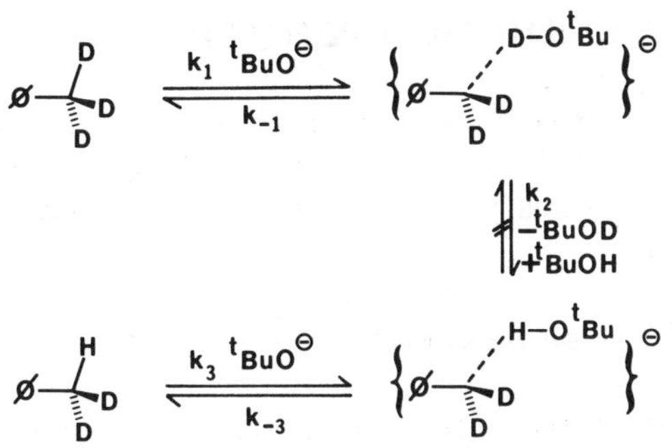

FIGURE 5.5.   Mechanism for H/D exchange in toluene-α, α, α-d_3.

Toluene-α,α,α-d_3 can be converted to unlabeled toluene by treatment with a catalytic amount of potassium *tert*-butoxide in *tert*-butanol (t-BuOH) or in dimethyl sulfoxide (DMSO). In accord with the hypothesis that we propose to test, one could imagine the mechanism shown in Figure 5.5.

In the case where DMSO is the solvent t-BuOH would be continuously regenerated by exchange of the acidic methyl hydrogens in CH_3SOCH_3. With either solvent the step labeled k_2 can be treated as essentially irreversible because of the large excess of protiated over deuterated solvent molecules. One can deduce the differential rate equation for this mechanism by applying the steady state approximation to the two hydrogen-bonded, anionic complexes.

$$\frac{-d[PhCD_3]}{dt} = \frac{k_1 k_2}{k_{-1} + k_2} [PhCD_3][t\text{-BuO}^-]$$

This can be reduced to two limiting cases:

Case I: If $k_2 \gg k_{-1}$ then $k_{obs} \simeq k_1$

Case II: If $k_2 \ll k_{-1}$ then $k_{obs} \simeq \dfrac{k_1}{k_{-1}} \cdot k_2$

Let us now consider th isotope effect on each step of the mechanism. The largest effects should occur where the changes in force constants are greatest, that is, k_1 and k_{-1} but not k_2. We would thus expect to see a large isotope effect on k_{obs} for Case I. For Case II, on the other hand, the isotope effects on k_1 and k_{-1} will tend to cancel and the observed effect should be quite small. Noting that k_2 is really a pseudo-first-order rate constant ($k_2 = k_2' [t\text{-BuOH}]$) we can anticipate that, of the two solvents, t-BuOH is more likely to exhibit Case I behavior. In DMSO, $[t\text{-BuOH}]$ is quite small and so Case II behavior is more probable.

Hofmann and coworkers[11] found $k_H/k_D = 0.62 \pm 0.15$ in DMSO at 32°C. Earlier work had shown that $k_H/k_D = 6.0$ in t-BuOH at 25°C. It would be difficult to explain the solvent dependence of the isotope effect without invoking the hydrogen bonded complexes that these workers were looking for.

5.3.   SECONDARY KINETIC ISOTOPE EFFECTS

A secondary isotope effect is one that occurs when a bond to the isotopically labeled atom is *not* broken in the transition state. These secondary effects are so much smaller than the primary ones that they are usually only measurable for the isotopes of hydrogen.

5.3.1.   α-Secondary Isotope Effects

α-Secondary isotope effects occur when deuterium or tritium is attached to a center that is undergoing a hybridization change. For example the acetolysis of 2-bromopropane-2-*d*:

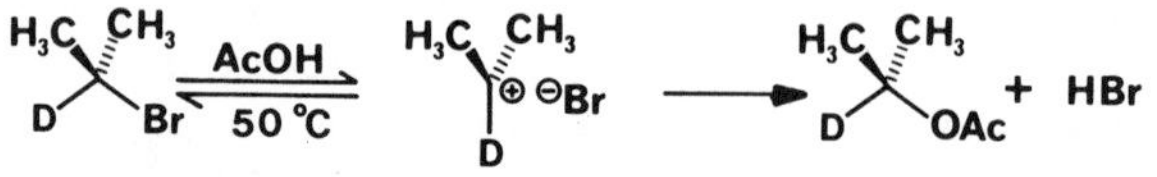

exhibits $k_H/k_D = 1.15$ at 50°C. The origin of this effect was first proposed by Streitwieser[12] in 1958. He took the empirical observation that $\tilde{\nu}_D \simeq \tilde{\nu}_H/1.35$ for C—D vs. C—H (the diatomic harmonic oscillator model would lead one to expect a factor of 1/1.36) and substituted into Eq. (2), obtaining

$$\frac{k_H}{k_D} \simeq \exp\left\{\frac{0.1865}{T}\sum_i (\tilde{\nu}_{Hi} - \tilde{\nu}_{Hi}^{\ddagger})\right\} \qquad (4)$$

The summation is over the (somewhat ill-defined) number of vibrations that involve significant contribution from the bond to the hydrogen that will be isotopically labeled.

For the specific example of 2-bromopropane one can identify three such vibrations: a stretch at 2890 cm^{-1} and two bending modes at 1340 cm^{-1}. Streitwieser estimated the corresponding frequencies in the carbonium ion by using acetaldehyde as a model. It was chosen because it has a hydrogen attached to a sp^2 hybridized carbon bearing a partial positive charge. The aldehyde hydrogen has a stretching vibration at 2800 cm^{-1}, an in-plane bend at 1350 cm^{-1}, and an out-of-plane bend at 800 cm^{-1}.

Clearly the largest change is in the vibration that becomes the out-of-plane bending mode. The change is such that the exponent of Eq. (4) will be positive, making $k_H/k_D > 1$. The Streitwieser model thus predicts that α-*secondary isotope effects should be > 1 (normal) for $sp^3 \to sp^2$ hybridization changes but < 1 (inverse) for $sp^2 \to sp^3$.* This prediction is borne out experimentally.

The quantitative predictions from Streitwieser's model are not quite so good. The frequencies listed above lead to a calculated value of $k_H/k_D = 1.43$ at 50°C. Part of the discrepancy with the experimental value of 1.15 is probably due to the poor model used for the carbonium ion but part is also probably due to solvent participation. If a solvent molecule attacked the secondary carbon as the bromide was leaving the transition state would resemble a trigonal bipyramid more than a trigonal planar geometry:

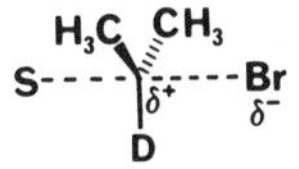

The presence of the two "axial ligands" would tend to increase the force constant for the out-of-plane vibration and decrease the magnitude of k_H/k_D. Support for this hypothesis comes from the observation that $k_H/k_D = 1.08$ in ethanol, which is presumably more nucleophilic than acetic acid. In fact the decrease in magnitude of an α-secondary isotope effect can be used to detect nucleophilic participation in solvolysis reactions, provided that one has some calibration standard for the isotope effect without participation.[13]

The parameter z that we used in the discussion of cyclopropane stereo-mutation (Sections 4.5.2 and 4.5.3) can now be recognized as an α-secondary isotope effect.

Summarizing, the α-secondary isotope effect can be used to detect changes in hybridization that occur during or before the rate-determining step of a reaction. Given suitable comparison data, it can also be used to detect nucleophilic participation in solvolysis reactions.

5.3.2. β-Secondary Isotope Effects

From our discussion so far one might expect that the magnitude of an isotope effect ought to decrease dramatically as the site of isotopic substitution is removed further from the reaction center. It turns out that this expectation is generally in accord with experimental fact, with some notable exceptions. Perhaps the most striking is the β-secondary isotope effect which, under certain circumstances, can be larger than an α-isotope effect. For example,

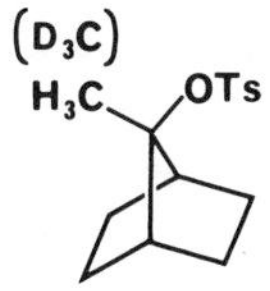

$$\frac{k_H}{k_D} = 1.89 \quad \text{(Solvolysis)}$$

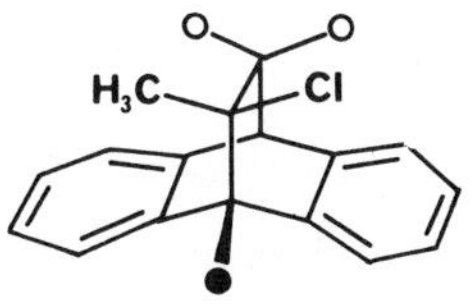

$$\frac{k_H}{k_D} = 1.14 \text{ per D at O}$$

$$\frac{k_H}{k_D} = 0.99 \text{ at } \bullet$$

The second of these examples comes from the work of Shiner[14] and suggests that the magnitude of the effect might depend on the dihedral angle between the leaving group and the β C—H bond at which the isotopic substitution is made.

More convincing evidence that this is indeed the case comes from later work of Shiner's group on the solvolysis of *tert*-butyl chloride-d_0, -d_1, -1,1-d_2, and -1,1,1-d_3.[15]

Let us assume that the solvolysis occurs predominantly from the staggered rotamer **6** of *tert*-butyl chloride and that it has an intrinsic rate constant k_H. In that case one could conceive of two different rate constants, k_g and k_t for the solvolyses of *gauche* and *trans* label isomers of **6-*d*₁**:

6–d_1 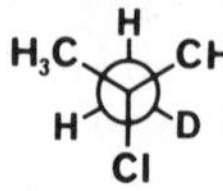

gauche

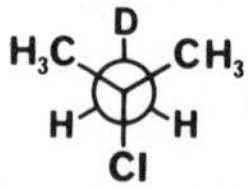

trans

If we define the mechanistic isotope effects z_g and z_t as $k_H/(3k_g)$ and $k_H/(3k_t)$, respectively, then we can express the expected phenomenological isotope effect, k_H/k_{D1}, as

$$\frac{k_H}{k_{D1}} = \frac{k_H}{k_t + 2k_g} = \frac{3z_g z_t}{z_g + 2z_t}$$

The experimental value for k_H/k_{D1} was 1.092.

For **6-d_2** there are also two label isomers, the *gauche–trans* and the *gauche–gauche* to which we will assign the solvolysis rate constants k_{gt} and k_{gg}, respectively.

6–d_2 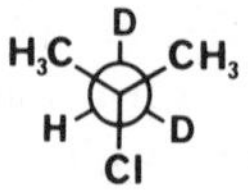

gauche-trans

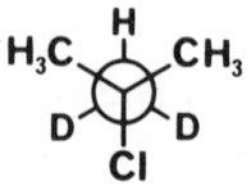

gauche-gauche

Assuming that replacement of hydrogen by deuterium at a particular site changes the rate constant by the same factor, regardless of whether the other sites carry H or D, we can write

$$\frac{k_H}{k_{D2}} = \frac{k_H}{2k_{gt} + k_{gg}}$$

$$= \frac{3z_g^2 z_t}{2z_g + z_t}$$

$$= 1.202 \text{ (experimental value)}$$

From these two experiments one has sufficient data to solve for the values of z_g and z_t. The results are

$$z_g = 1.004$$

$$z_t = 1.325$$

If the assumptions involved in this analysis are valid and the data have sufficient precision, these values show a clear dihedral angle dependence of the β-secondary isotope effect. Had there been none one would have found $z_g = z_t$. Both the assumptions and the precision can be tested by using the values of z_g and z_t to predict the value of k_H/k_{D3} for *tert*-butyl chloride-1,1,1-d_3. The calculated value is 1.336 (i.e., $z_g^2 z_t$). The experimental value[15] was 1.330.

The dihedral angle dependence is consistent with a hyperconjugation model for the β-secondary isotope effect. Remembering that the heavier isotope prefers to be at the more strongly bound site, we can recognize that a C—D bond would be less able than a C—H bond to participate in hyperconjugation. This would lead to a kinetic isotope effect that was >1, with a maximum value when the C—H(D) bond was coplanar with the bond to the leaving group, in accord with experimental observation. Since hyperconjugation can only occur for a β C—H bond, this model also explains why the β-secondary isotope effect is abnormally large when compared to α and γ effects.

To date there has been little deliberate use of the β-secondary isotope effect to distinguish among mechanistic hypotheses.[15] It is more a factor that must be taken into account when one analyzes H/D isotope effects, especially in reactions involving carbonium ions. Baldwin[16] has suggested that certain biradical reactions might also exhibit large β-isotope effects.

5.3.3. Steric Isotope Effects

Remote isotopic substitution (i.e., at a site that is γ related or further from the reaction site) generally has an effect that is so small as to be negligible. The major exception is when the site of substitution is severely sterically encumbered in the reactant or becomes so in the transition state. Under these circumstances one can sometimes see a so-called steric isotope effect.

There are two contributions to the steric isotope effect, both resulting in a C—D bond appearing to be shorter than the corresponding C—H bond. The first is that, in an anharmonic potential function, the center of the zero point vibrational level (in other words the maximum in the vibrational probability distribution) is displaced toward greater R as the energy of that level is increased (see Figure 5.6). Since all vibrations involving C—H bonds have higher zero point energy than those involving corresponding C—D bonds, the result is that the most probable bond distance is slightly larger for C—H than for C—D. This

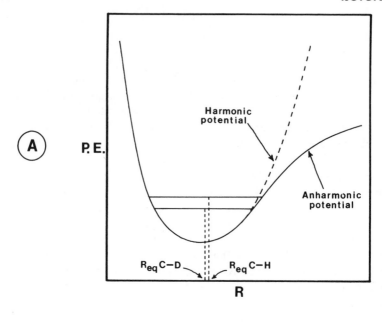

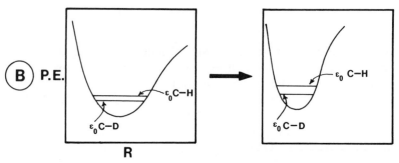

FIGURE 5.6. Two contributions to the steric isotope effect. In A the anharmonicity of the C—H(D) stretching vibration causes C—H to have a slightly greater equilibrium length than C—D. In B the increase in force constants that accompanies steric congestion increases the energetic advantage for C—D over C—H.

is supported by analyses of the IR and Raman spectra of CH_4 and CD_4, which show that the C—H distance is longer than the C—D distance by 0.001 Å.[17]

The second contribution is that force constants are generally increased in sterically crowded situations, resulting in an increase in the zero point energy difference between isotopically related molecules (Figure 5.6).

The two contributions combine to make D more "comfortable" than H at a sterically encumbered site. The resulting kinetic isotope effect can thus be normal or inverse depending on the degree of steric hindrance in the reactant and the transition state:

$$\frac{k_{H\text{-}}}{k_D} = 0.86$$

$$\frac{k_{H\text{-}}}{k_D} = 1.107$$

Again, there has not yet been much use of the steric isotope effect as a tool for elucidating reaction mechanisms but knowledge of its existence is necessary if other isotope effects are to be interpreted properly.

5.4. Inter- vs. Intramolecular Isotope Effects

In a molecule containing two or more symmetry-related reaction sites, one can measure two different types of isotope effect. The first is the normal type of intermolecular isotope effect in which the rate constants for reaction of unlabeled and labeled molecules are compared. The second is an intramolecular competition in which only some of the symmetry-related reaction sites are labeled and the isotope effect is measured by analysis of the label distribution in the product(s). Identity of these two isotope effects provides good evidence that the rate-determining and product-determining steps of the reaction are the same (unless both isotope effects are found to be unity, in which case no conclusion can be drawn). Nonidentity provides evidence for the existence of an intermediate, although, as we shall see, caution is required in the interpretation of such a result.

One of the first examples of this type of mechanistic analysis was an elegant study of allene dimerization by Dolbier.[18] He was able to measure both inter- and intramolecular isotope effects (which in this case were of the α-secondary type) by analyzing product distributions. The intermolecular effect was determined by codimerizing equal proportions of C_3H_4 and C_3D_4. The three label-isomeric dimethylenecyclobutanes were formed within experimental error of their statistical ratios, indicating the absence of an isotope effect on the rate determining step.

$C_3H_4 + C_3D_4 \longrightarrow$

50.7% 49.3%

24.1   49.9   26.0

$\bullet = CD_2$

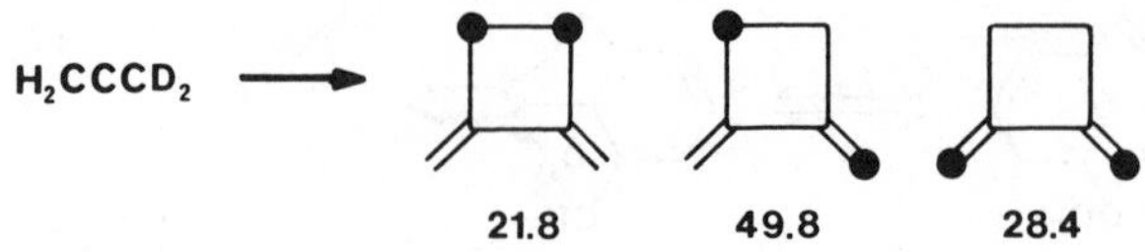

$$H_2CCCD_2 \longrightarrow$$

21.8   49.8   28.4

The intramolecular isotope effect was determined by dimerizing allene-1,1-d_2. In this case the product ratios deviated significantly from their statistical values, indicating that there was an isotope effect on the product determining step. The ratios are consistent with a value of $k_H/k_D = 0.936$ per deuterium.

Dolbier's results are clearly inconsistent with a $_\pi 2_s + _\pi 2_a$ pericyclic mechanism which, having only a single step, would require that inter- and intramolecular isotope effects be identical. The existence of an intermediate is thus strongly implicated. Apparently this intermediate is formed without significant rehybridization at the methylene carbons since there is no detectable isotope effect on the rate-determining step. A plausible candidate that fits this criterion is the tetramethylene ethane biradical 7. Note that if this biradical is the

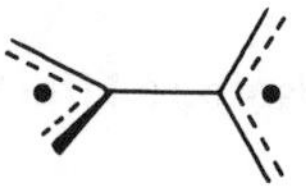

intermediate one can say that it must either have the "orthogonal" (D_{2d}) geometry depicted or must be capable of undergoing rotation about the central C—C bond at a rate comparable to or greater than the rate of ring closure. Were neither of these conditions true the label distribution in the products of $C_3H_2D_2$ dimerization would again be statistical.

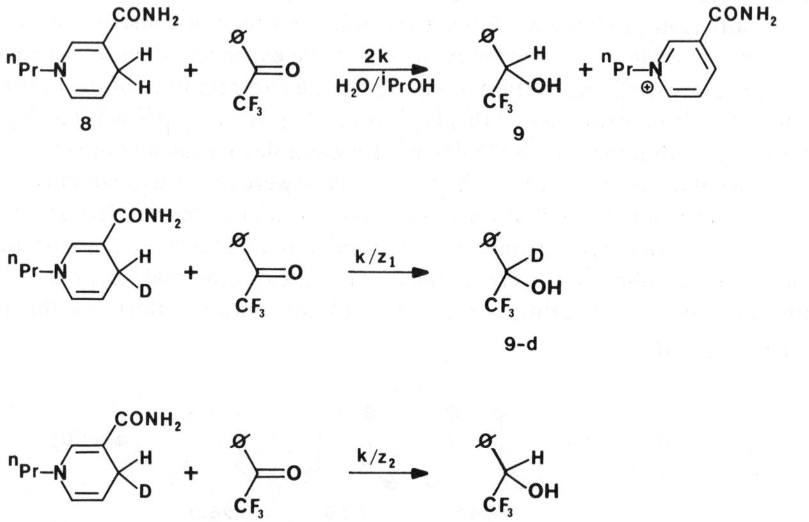

FIGURE 5.7.   Direct hydride transfer mechanism for the reduction of trifluoroacetophenone by 8.

The interpretation of Dolbier's results is quite straightforward and unambiguous because of the simplicity of the reaction. However there is need for caution when applying this technique to more complex reactions, as illustrated by the study of Chipman and coworkers[19] on the NADH model **8** (see Figure 5.7). **8** will reduce electrophilic ketones such as trifluoroacetophenone at 50° C in a medium of aqueous isopropanol. The simplest mechanism for this reaction would be the direct hydride transfer depicted in Figure 5.7. If this mechanism were correct, one would expect inter- and intramolecular isotope effects to be identical.

The experimental results were that $k_H/k_D = 1.16 \pm 0.04$ at 50° C and [9]/[9-d] $= 3.8 \pm 0.3$ at 50° C. Some manipulation of these data is required before the mechanistic isotope effects can be determined.

As shown in Figure 5.7, one must consider two types of mechanistic isotope effect: a primary type (z_1) when D is transferred and an α-secondary type (z_2) when D remains behind. If k represents the mechanistic rate constant for transfer of one hydrogen from **8** then we can write

$$\frac{k_H}{k_D} = \frac{2k}{k/z_1 + k/z_2}$$

$$= \frac{2z_1 z_2}{z_1 + z_2}$$

$$= 1.16$$

$$\frac{[9]}{[9\text{-}d]} = \frac{z_1}{z_2}$$

$$= 3.8$$

The authors assumed that $z_2 \simeq 1$ in which case z_1 (intermolecular) $= 1.38 \pm 0.11$ whereas z_1 (intramolecular) $= 3.8 \pm 0.3$. Given the validity of the assumption about the magnitude of z_2, these results would appear to rule out the direct hydride transfer.

An alternative way of analyzing the data is to solve for the values of z_1 and z_2 that would be necessary to explain the experimental results. This approach gives $z_1 = 2.78$ and $z_2 = 0.733$. The value for the α-secondary isotope effect looks implausible because it is substantial and inverse—for a $sp^3 \rightarrow sp^2$ hybridization change one would expect a normal secondary isotope effect.

Thus the mechanism depicted in Figure 5.7 appears to be inconsistent with the experimental isotope effects however one analyzes the data. In view of the discussion at the beginning of this section one might be tempted to revise the mechanism by including an intermediate between the reactants and products. In fact the data *do not* require this. An intermediate *is* required but *it need not be on the path from reactants to products*. In the present example, further investigation by Chipman and coworkers[20] revealed the reversible formation of a side product (**11**):

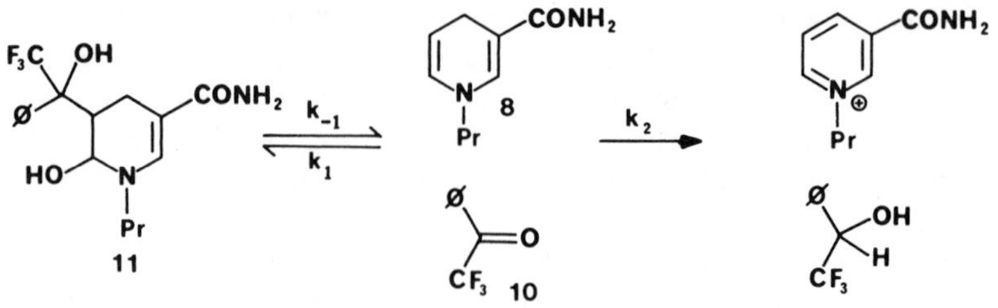

$$\frac{-d[\mathbf{8}]}{dt} = (k_1 + k_2)[\mathbf{8}][\mathbf{10}] - k_{-1}[\mathbf{11}]$$

The rate constants k_1 and k_{-1} serve to "dilute" the isotope effect on k_2 when it is determined by the intermolecular method. The intramolecular competition gives the true isotope effect on k_2. Numerical integration of the rate equation allowed the determination of the true intermolecular isotope effect, which was found to be identical to the intramolecular effect within experimental error.

This result does not prove that direct hydride transfer is necessarily the correct mechanism. The data would permit formation of an intermediate in a fast preequilibrium prior to hydrogen transfer. All one can say with confidence is that the products are formed in the rate-determining step.

5.5. THE INDUCED KINETIC ISOTOPE EFFECT

Consider the following generalized mechanism:

$$\begin{array}{ccccccccc} B & \xleftarrow{k_2} & I & \underset{k_1}{\overset{k_{-1}}{\rightleftharpoons}} & A & \underset{k_{-1}}{\overset{k_1}{\rightleftharpoons}} & I' & \xrightarrow{k_2} & B' \\ & & {\scriptstyle k_3}\downarrow & & & & \downarrow{\scriptstyle k_3/z} & & \\ & & C & & & & C' & & \end{array}$$

The starting material in this scheme is A. I is a steady-state intermediate. I′, B′, and C′ are label-isomers of I, B, and C, respectively.

Mechanisms of this type are often very difficult to distinguish from alternatives in which the two label isomers of each product are formed from the same intermediate:

$$\begin{array}{ccccccccc} B & \xleftarrow{k_3} & I & \underset{k_1}{\overset{k_{-1}}{\rightleftharpoons}} & A & \underset{k_{-2}}{\overset{k_2}{\rightleftharpoons}} & J & \xrightarrow{k_4} & C \\ & & {\scriptstyle k_3}\downarrow & & & & \downarrow{\scriptstyle k_4/z} & & \\ & & B' & & & & C' & & \end{array}$$

The induced kinetic isotope effect (IKIE) provides a means of making such a distinction.[21]

Applying the steady-state approximation to I and I' in the first mechanism one finds:

$$\frac{[B']}{[B]} = \frac{k_{-1} + k_2 + k_3}{k_{-1} + k_2 + k_3/z}$$

$$> 1 \qquad \text{if } z > 1$$

The isotope effect on the steps $I \to C$ and $I' \to C'$ causes a different fractionation of the two steady-state intermediates and induces an apparent isotope effect on the product ratio $[B']/[B]$ despite the fact that $I \to B$ and $I' \to B'$ have identical rate constants.

In contrast, the second mechanism predicts $[B']/[B] = 1$ regardless of the value of z. This situation obtains even if B, B', C, and C' are all formed from a single intermediate.

The utility of this analysis is illustrated by the work of Samuelson[21] who studied the reaction of *trans*-2,3-diphenylmethylenecyclopropane (**12**) with diiron nonacarbonyl. The stereochemistry of the trimethylenemethane complex (**13**) indicated a disrotatory ring opening mode.

Two mechanisms were considered for the formation of **13** (see Figure 5.8). In the first (A) an $Fe(CO)_4$ intermediate was proposed to act as an electrophile and to attack the methylenecyclopropane, forming a zwitterionic intermediate. The cationic part of this zwitterion would then undergo an electrocyclic ring opening, which should be disrotatory according to the Woodward–Hoffmann rules.

In the second mechanism (B) the $Fe(CO)_4$ was proposed to form a π complex with **12** and then to lose CO prior to undergoing the ring opening reaction. Qualitative molecular orbital arguments suggested that the disrotatory ring opening should occur with bending of the C—C σ bond away from the metal.[22]

Earlier work[23] had implicated a methylenecyclopropane $Fe(CO)_3$ complex as an intermediate in the formation of the butadiene complex:

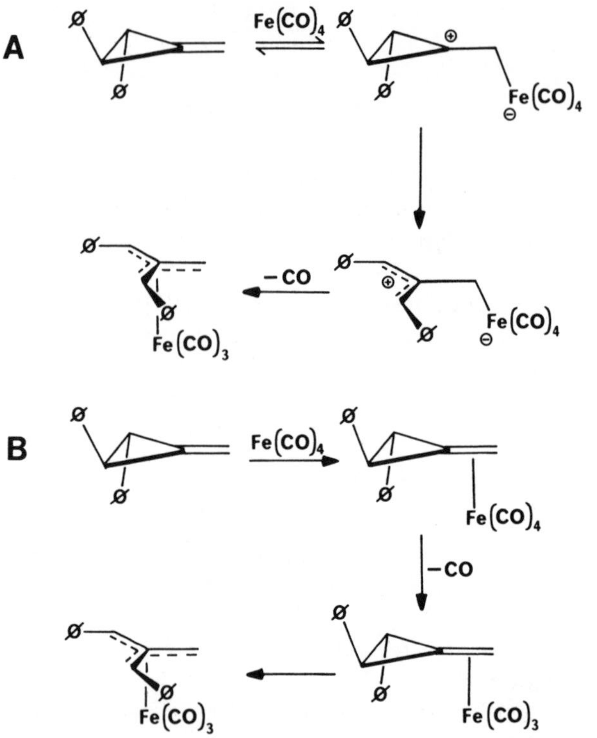

FIGURE 5.8. Mechanism for the reaction of racemic 2,3-diphenylmethylenecyclopropane with Fe₂(CO)₉.

and so if mechanism B were correct, use of the deuterated reactant **12-d** should result in observation of an IKIE on the product ratio **13'-d** : **13-d** (see Figure 5.9). In contrast mechanism A would predict that **13'-d** : **13-d** should differ from unity only to the extent of a possible steric isotope effect. The experimental result was that **13'-d** : **13-d** = 1.22 ± 0.03. In order to determine the magnitude of the steric isotope effect the reaction was also run with 2,2-diphenylmethylenecyclo-propane-3-*d* (**14-d**). This substrate was known to give none of the butadiene product and so would be unable to exhibit IKIE regardless of the mechanism of formation of the trimethylenemethane complex. The steric isotope effect, on the other hand, should be virtually identical for the two methylenecyclopropanes.

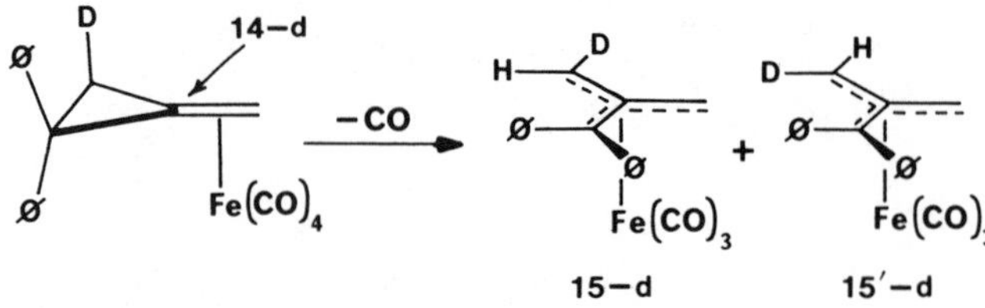

The result was that **14-d** gave **15'-d** and **15-d** in a ratio of (1.02 ± 0.02) : 1, indicating that the isotope effect on the ratio of **13'-d** : **13-d** was due largely to

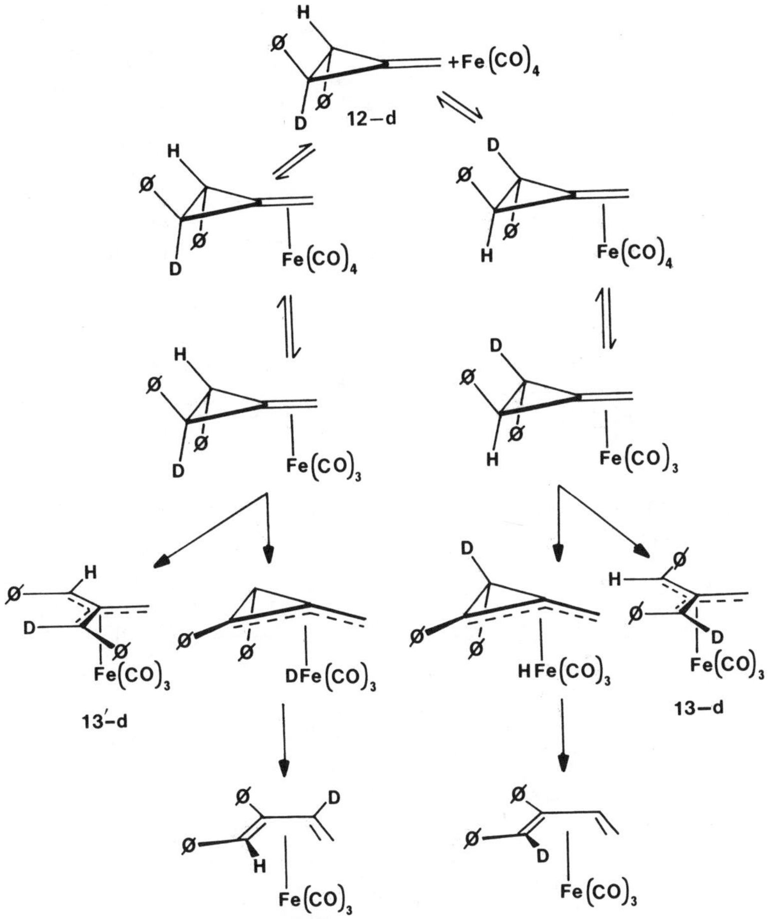

FIGURE 5.9. Application of mechanism B to racemic 2,3-diphenylmethylenecyclopropane-2-*d*.

IKIE. This, in turn, provides support for mechanism B and rules out mechanism A.

5.6. THORNTON ANALYSIS

The Thornton analysis[24] is an attempt to deduce something about the symmetries of transition states by measurement of isotope effects. It is very similar in philosophy to the Tolbert analysis that we discussed in Section 3.5. As an illustration of the technique the thermal fragmentation of **16** to anthracene and ethylene will be considered. As before, two mechanisms seem plausible for this reaction: the concerted retro-Diels-Alder process (A) and the biradical reaction (B). For the sake of simplicity we will focus on the cyclohexene portion of **16**.

The experiment consists of measuring the rate constants for fragmentation of

16

16, **16-d_2**, and **16-d_4**. If we call the secondary isotope effect due to one CD_2 unit z, then we can write

By calling the rate constant for reaction of **16-d_4** k/z^2 one is making the implicit assumption that there is no "isotope effect on an isotope effect," in other words there is a constant factor reducing the rate constant for each CD_2 that is introduced. A similar assumption was made in Shiner's analysis of the β-secondary isotope effect and could be shown to be valid in that case (Section 5.3.2). Given this assumption, mechanism A thus leads to a predicted relationship between the phenomenological isotope effects, namely,

$$\frac{k_H}{k_{d4}} = \left\{\frac{k_H}{k_{d2}}\right\}^2$$

The asymmetry of the biradical (or the transition state leading to it) in mechanism B leads to the creation of two different sites for CD_2 substitution. One must therefore consider two mechanistic isotope effects that we will call z_α and z_β. The mechanistic rate constants can be written from which one deduces the phenomenological isotope effects

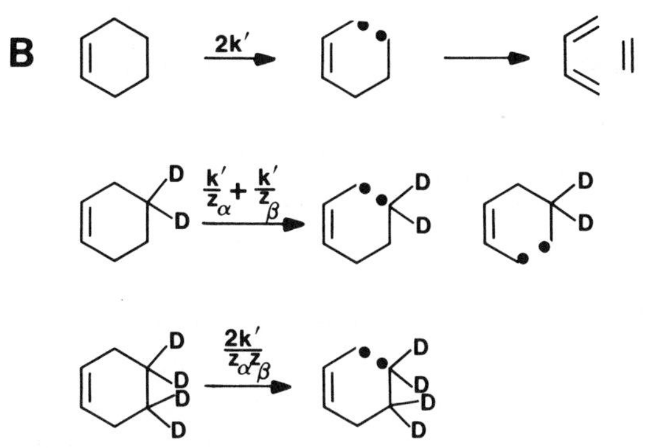

$$\frac{k_H}{k_{d2}} = \frac{2z_\alpha z_\beta}{z_\alpha + z_\beta}$$

$$\frac{k_H}{k_{d4}} = z_\alpha z_\beta$$

Hence,

$$\frac{k_H}{k_{d4}} \neq \left\{\frac{k_H}{k_{d2}}\right\}^2 \qquad \text{unless } z_\alpha = z_\beta$$

The experimental results were

$$\left\{\frac{k_H}{k_{d2}}\right\}^2 = 1.37$$

$$\frac{k_H}{k_{d4}} = 1.36$$

These were taken to be equal within experimental error and hence support for mechanism A over mechanism B was claimed.[24]

Unfortunately the Thornton analysis suffers from the same lack of calibration that was apparent in the Tolbert analysis. How large a deviation would one expect if mechanism B were operating? In the present case an estimate can be made.

Both z_α and z_β would be expected to have lower limits of 1.00. In order to explain the observed value for k_H/k_{d2} by mechanism B one would therefore have to place an upper limit of 1.41 on both of these mechanistic isotope effects. The upper limit that could be expected for k_H/k_{d4} is thus 1.41, which seems perilously close to the observed value of 1.36. If both z_α and z_β were > 1, as one might reasonably expect, then the calculated value for k_H/k_{d4} would be even closer to the observed one. It appears, then, that the method suffers from a lack of sensitivity, at least for this example. There could well be other mechanistic problems that would be better suited to this type of analysis.

5.7. SOLVENT ISOTOPE EFFECTS

Studies on the effect of changing solvent isotopic composition have, to date, been restricted almost entirely to the pair H_2O/D_2O, although similar investigations could presumably be carried out in other hydroxylic solvents.

The reactions are usually acid or base catalyzed and the mechanistic problem is to determine whether proton transfer is rate determining or occurs in a fast preequilibrium step. Gold[25] and Kresge[26] have developed detailed models that allow one to predict the effect of changing the solvent from H_2O to D_2O on the rates of a wide variety of such reactions. In this section we will consider a much simpler model that provides less detailed information but, perhaps, gives a clearer physical picture of the factors affecting the reaction rate.

If the proton transfer is rate determining then the situation is quite straightforward: replacement of H_2O by D_2O will cause a reduction in rate provided that the transferred hydrogen has been replaced by deuterium. Under these circumstances the solvent change can be viewed as constituting an *in situ* synthesis of an isotopically substituted reactant, and the reduction in rate is then just the result of a normal primary kinetic isotope effect.

Fast, reversible proton transfer prior to the rate-determining step leads to a somewhat more complex situation. Consider the reaction

$$S + H_3O^+ \overset{K}{\rightleftharpoons} SH^+ + H_2O$$

$$SH^+ \overset{k}{\longrightarrow} \text{Products}$$

If the second step is rate determining then

$$\frac{-d[S]}{dt} = Kk[S][H_3O^+]$$

Usually k is little affected by the substitution of D_3O^+ for H_3O^+ since the bond(s) broken in the rate-determining step · rely involve(s) the newly added hydrogen ion. At most one would expect a secondary kinetic isotope effect. On the other hand K is substantially affected by isotopic substitution. When SH^+ is a weaker acid than H_3O^+ (i.e., $K > 1$), the $S—H^+$ bond is presumably stronger than the $H_2O—H^+$ bond (although in reality there will be extensive hydrogen bonding of both acids to the solvent, which confuses the issue somewhat). Remembering that the heavier isotope prefers to be in the more strongly bound site, we can anticipate that replacement of H_2O by D_2O will increase K and thus increase the rate of the reaction. This is indeed the observed result for most reactions involving rapid prequilibrium proton transfer. Effects are typically in the range of 25–100% at 25° C.[27] Note, however, that one could conceive of situations in which the equilibrium isotope effect would favor $H_3O^+(D_3O^+)$ over $SH^+(SD^+)$, especially if $K < 1$. Under such circumstances the replacement of H_2O by D_2O would decrease the reaction rate. Thus solvent isotope effects cannot be used as rigorous means of determining reaction mechanism in the absence of detailed information about equilibrium constants. Their measurement provides one piece of information that contributes to the development of a mechanistic hypothesis. Other ways of determining the timing of proton transfers will be considered in Chapter 6.

REFERENCES

1. C. J. H. Schutte, *The Wave Mechanics of Atoms, Molecules and Ions.* Arnold, London, 1968.

2. W. J. Moore, *Physical Chemistry*, 4th ed., Prentice-Hall, Englewood Cliffs, N.J., 1972, p. 381.

3. R. B. Timmons, J. De Guzman, and R. E. Varnerin, *J. Am. Chem. Soc.*, **90**, 5996 (1968) and references therein.

4. F. H. Westheimer, *Chem. Rev.*, **61**, 265 (1961).

5. M. J. Stern and M. Wolfsberg, *J. Chem. Phys.*, **45**, 4105 (1966).

6. L. Melander and W. H. Saunders, Jr., *Reaction Rates of Isotopic Molecules*, Wiley-Interscience, New York, 1980.

7. G. Bunton, D. Griller, L. R. C. Barclay, and K. U. Ingold, *J. Am. Chem. Soc.*, **98**, 6803 (1976).

8. E. Wigner, *Z. Phys. Chem. B*, **19**, 203 (1932).

9. B. W. Palmer and A. Fry, *J. Am. Chem. Soc.*, **92**, 2580 (1970).

10. D. J. Cram, F. Willey, and H. P. Fischer, H. M. Relles, and D. A. Scott, *J. Am. Chem. Soc.*, **88**, 2759 (1966).

11. J. E. Hofmann, A. Schriesheim, and R. E. Nickols, *Tetrahedron Lett.*, 1745 (1965).

12. A. Streitwieser, Jr., R. H. Jagow, R. C. Fahey, and S. Suzuki, *J. Am. Chem. Soc.*, **80**, 2326 (1958).

13. V. J. Shiner, Jr., in *Isotope Effects in Chemical Reactions*, C. J. Collins and N. S. Bowman (Eds.), Van Nostrand Reinhold, New York, 1970.

14. V. J. Shiner, Jr. and J. S. Humphrey, *J. Am. Chem. Soc.* **85**, 2416 (1963).

15. V. J. Shiner, Jr., B. L. Murr, and G. Heinemann, *J. Am. Chem. Soc.*, **85**, 2413 (1963).

16. J. E. Baldwin and C. G. Carter, *J. Am. Chem. Soc.*, **101**, 1325 (1979).

17. D. P. Stevenson and J. A. Ibers, *J. Chem. Phys.*, **33**, 762 (1960).

18. S. -H. Dai and W. R. Dolbier, Jr., *J. Am. Chem. Soc.*, **94**, 3946 (1972).

19. J. J. Steffens and D. Chipman, *J. Am. Chem. Soc.*, **93**, 6649 (1971).

20. D. M. Chipman, R. Yaniv, and P. van Eikeren, *J. Am. Chem. Soc.*, **102**, 3244 (1980).

21. A. G. Samuelson and B. K. Carpenter, *J. Chem. Soc. Chem. Commun.*, 354 (1981). A similar analysis that predates the work of Samuelson and Carpenter can be found in: T. J. Katz and S. A. Cerefice, *J. Am. Chem. Soc.*, **91**, 6519 (1969).

22. A. R. Pinhas and B. K. Carpenter, *J. Chem. Soc. Chem. Commun.*, 15 (1980).

23. A. R. Pinhas, A. G. Samuelson, R. Risemberg, E. V. Arnold, J. Clardy, and B. K. Carpenter, *J. Am. Chem. Soc.*, **103**, 1668 (1981).

24. M. Taagepara and E. R. Thornton, *J. Am. Chem. Soc.*, **94**, 1168 (1972).

25. V. Gold, *Trans. Faraday Soc.*, **56**, 255 (1960).

26. A. J. Kresge, *Pure Appl. Chem.*, **8**, 243 (1964).

27. See Chapter 7 of reference 6.

CHAPTER 6

MECHANISMS OF ACID–BASE REACTIONS

Reactions involving the transfer of a proton form a special class for which unique experimental techniques have been developed.

There are a number of reasons why the mechanisms of these reactions are particularly challenging to investigate. Proton transfers are usually very fast, often diffusion controlled, and the number of functional groups capable of donating or accepting a proton is very large. Often there will be several species in a given solution that are capable of participating in proton transfer reactions and so the kinetics can become quite complex. Many acid–base reactions are run in an aqueous medium where participation by the solvent is frequently important.

All of these factors tend to vitiate the kinds of experiments that we have discussed in the preceeding chapters, where single sites have been isotopically labeled or single reactants made optically active. Nevertheless the importance of acid–base reactions, not least in biological systems, has led to a considerable amount of research into their mechanisms. In this chapter we will summarize some of the techniques and concepts that have resulted from this effort.

6.1. ACIDITY FUNCTIONS

All of our discussions so far have implicitly assumed that the solutions under consideration exhibit ideal behavior. Thus, for example, our analyses of kinetic problems have used concentrations rather than activities of the solutes. For most purposes the assumption of ideality causes no problems, however concentrated aqueous solutions of mineral acids deviate so severely from ideal behavior that some correction is necessary.

The definition of pH, being $-\log_{10} a_{H^+}$, requires some estimate of activity coefficients since the free hydrogen ion activity is not directly measurable. The methods used to calculate the activity coefficients[1] are not reliable above ionic strengths of about 0.1 m and so are clearly not applicable to concentrated aqueous acids. Since many acid-catalyzed reactions require a medium of this kind, it becomes desirable to find some parameter that can describe the proton donating ability of the solution.

Hammett and Deyrup[2] attacked the problem by studying the protonation of various substituted aniline bases. Their work led them to propose a so called "acidity function," H_0, that served the purpose outlined above.

For an equilibrium such as

$$B + H_3O^+ \rightleftharpoons BH^+ + H_2O$$

where B is some (uncharged) base one can write

$$K_{BH^+} = \frac{[BH^+]\gamma_{BH^+}}{[B]\gamma_B a_{H_3O^+}}$$

In this equation $a_{H_3O^+}$ is the activity of available acid and γ_i is the activity coefficient of species i. Hence,

$$pH = pK_{BH^+} - \log\left(\frac{[BH^+]}{[B]}\right) - \log\left(\frac{\gamma_{BH^+}}{\gamma_B}\right)$$

Thus one could hope to determine the dissociation constant of the protonated base in dilute solution and then use the measured ratio $[BH^+]/[B]$, perhaps determined spectrophotometrically, to calculate the pH of a new solution. For dilute solutions the ratio of activity coefficients could be estimated from Debye-Hückel theory or one of its modifications.[3] Unfortunately, for concentrated solutions there is no model that allows accurate determination of the activity coefficients and so the proton donating ability of the medium can no longer be determined in this way. Hammett and Deyrup sidestepped this problem by defining an acidity function in which the troublesome activity coefficients were simply neglected:

$$H_0 = pK_{BH^+} - \log\left(\frac{[BH^+]}{[B]}\right)$$

$$= pH + \log\left(\frac{\gamma_{BH^+}}{\gamma_B}\right)$$

At infinite dilution the activity coefficients would become unity and H_0 would become identical with pH. At higher concentrations one could hope that the ratio γ_{BH^+}/γ_B would be approximately constant for all neutral bases and that the H_0 would thus provide a useful measure of acidity, differing from pH only by some constant term. For the substituted anilines (often called Hammett indicators) the approximate constancy of γ_{BH^+}/γ_B was verified by using different bases to obtain independent estimates of H_0.

Unfortunately, larger changes in the structure of the base cause substantial deviations and lead to a much more restricted range of applicability than originally hoped.

The assumption of constant γ_{BH^+}/γ_B in Hammett and Deyrup's definition led to the development of a plethora of different acidity functions for bases of different structural type. Bunnet and Olsen attempted to improve this situation by modifying the model somewhat. Instead of assuming that γ_{BH^+}/γ_B should be constant, they assumed that the activity coefficient ratios would be linearly related for different bases.[4] Their treatment allowed a variety of structurally different bases to be treated with the same acidity function although a base-dependent term was now included.

Despite the limitations of Hammett's acidity function model and its later variants, useful mechanistic information can be gained by studying the variation in rate (or more precisely $\log k_{obs}$) of a reaction with the change in acidity of the medium. Early work in this area centered on differentiation between acid-catalyzed hydrolyses in which the rate-determining step was unimolecular (A-1 in the Ingold notation[5]) and those in which the rate-determining step was biomolecular (A-2), the latter typically involving nucleophilic attack by a water molecule.

A generalized A-1 mechanism can be written:

$$S + H_3O^+ \rightleftharpoons SH^+ + H_2O$$

$$SH^+ \xrightarrow{k} R^+$$

$$R^+ + H_2O \longrightarrow products$$

The rate-determining step is the second, with rate constant k. If one measures a first-order rate constant, k_{obs}, for the disappearance of total substrate (i.e., $S + SH^+$) then

$$k_{obs} = -\left\{\frac{1}{[S] + [SH^+]}\right\} \frac{d([S] + +SH^+])}{dt}$$

$$= -\left\{\frac{[SH^+]}{[S] + [SH^+]}\right\} k \frac{\gamma_{SH^+}}{\gamma^{\ddagger}}$$

$\gamma^{\ddagger}$ is the activity coefficient for the transition state of the rate-determining step. One can relate $[SH^+]$ to the dissociation constant K_{SH^+} and the H_0 acidity function, giving

$$\log k_{obs} = H_0 + \log\left(\frac{k}{K_{SH^+}}\right) + \frac{[S]}{[S] + [SH^+]} \cdot \frac{\gamma_{BH^+}\gamma_S}{\gamma_B\gamma^{\ddagger}}$$

Under normal circumstances $[S] \gg [SH^+]$ and so the concentration ratio in the third term is near unity. In their original derivation Zucker and Hammett[6] further assumed that the activity coefficient ratio γ_{BH^+}/γ_B would be approximately cancelled by the ratio $\gamma_S/\gamma^{\ddagger}$. With hindsight one can recognize that this is equivalent to the assumption that acidity functions would be independent of base structure, which we know to be false. The Zucker-Hammett postulate leads to the prediction that log k_{obs} vs. H_0 should be a linear plot with unit slope for A-1 type hydrolysis mechanisms. In practice one does find linear plots for reactions such as the acid-catalyzed hydrolyses of β-propiolactone,[7] acetaldehyde dimethyl acetal,[8] and epichlorhydrin[9] but the slopes often differ significantly from unity.

A typical A-2 mechanism would be

$$S + H_3O^+ \rightleftharpoons SH^+ + H_2O$$

$$SH^+ + H_2O \xrightarrow{k} \text{products}$$

Following the type of derivation outlined above, one finds in this case

$$\log k_{obs} = \log C_{H_3O^+} + \log\left\{\frac{k}{K_{SH^+}}\right\} + \log\left\{\frac{\gamma_S \gamma_{H_3O^+} a_w}{\gamma^{\ddagger}}\right\}$$

In this expression a_w is the activity of water which enters because of the involvement of a water molecule in the rate-determining step. $C_{H_3O^+}$ is the *primitive concentration* of acid, in other words the total number of moles of acid added to the solution, regardless of its degree of dissociation or the exact location of the protons. Terms in square brackets, by contrast, can be considered to be sophisticated concentrations, that is, the concentrations of individual molecular or ionic species.

For aqueous solutions the third term can be taken to be near zero, with the result that one expects a linear relationship between log k_{obs} and log $C_{H_3O^+}$. The hydrolyses of common esters such as methyl acetate and methyl benzoate exhibit such behavior.[10]

A third plausible mechanism for hydrolysis, rate-determining proton transfer, cannot be distinguished from the A-1 mechanism by the criteria considered here. Experimental techniques that do allow such a distinction will be considered in Section 6.3.

The Bunnett-Olsen variation on the formulation of acidity functions can be applied to the mechanistic problems discussed above[11] but will not be considered here.

6.2. pH-RATE PROFILES

The studies of acidity dependence of reaction rates described in Section 6.1 can, of course, be extended to a wider range of hydrogen ion activity. Frequently such extrapolations, loosely called pH–rate profiles despite the fact that pH is not

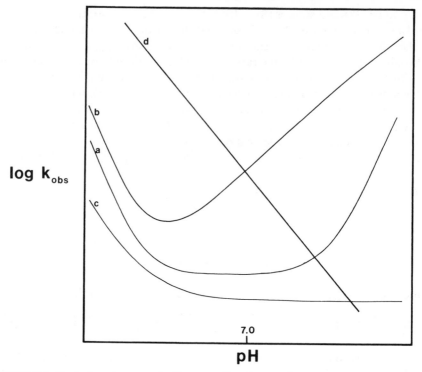

FIGURE 6.1. Typical appearances of pH rate profiles. See text for explanation of labels on the graphs.

always the appropriate measure of acidity, lead to sharp discontinuities in the previously observed linear plots. The turning points signal changes in mechanism (or changes in rate-determining step within the same mechanism).

Some commonly encountered shapes for these pH–rate profiles are illustrated qualitatively in Figure 6.1.

Line **a** with its two turning points and flat region around pH 7 is typical of a reaction such as the mutarotation of glucose, which is both acid and base catalyzed but which also has an appreciable uncatalyzed rate.

H₂COH ... HO HO HO C⊖OH → H₂COH O⊖ HO HO HO **Base Catalyzed**

In line **b** the central flat region is missing but the rate is increased at both high and low pH. This indicates a reaction that is acid and base catalyzed but which has no significant uncatalyzed rate. An example would be methyl acetate hydrolysis.

$$CH_3-C(\overset{\oplus}{O}H)-OCH_3 \;\underset{}{\overset{H_2O}{\rightleftharpoons}}\; CH_3-C(HO)(\overset{\oplus}{O}H_2)-OCH_3 \longrightarrow$$ **Acid Catalyzed**

$$CH_3-C(=O)-OCH_3 \;\underset{}{\overset{HO^{\ominus}}{\rightleftharpoons}}\; CH_3-C(\overset{\ominus}{O})(OH)-OCH_3 \longrightarrow$$ **Base Catalyzed**

Line **c** indicates acid catalysis and some uncatalyzed reaction but no base catalysis. Trimethyl orthoacetate hydrolysis would be an example.

$$CH_3-C(\overset{\oplus}{O}\overset{H}{CH_3})(OCH_3)(OCH_3) \rightleftharpoons CH_3-C(\overset{\oplus}{O}CH_3)(OCH_3) + HOCH_3 \rightleftharpoons CH_3-C(\overset{\oplus}{O}H_2)(OCH_3)(OCH_3) \longrightarrow$$ **Acid Catalyzed**

$$CH_3-C(OCH_3)(OCH_3)(OCH_3) \rightleftharpoons CH_3-C(\overset{\oplus}{O}CH_3)(OCH_3) + {}^{\ominus}OCH_3 \rightleftharpoons CH_3-C(\overset{\oplus}{O}H_2)(OCH_3)(OCH_3) \longrightarrow$$ **Uncatalyzed**

Line **d** is the typical plot found for acetal hydrolyses and indicates that only acid catalysis is effective.

$$CH_3-C(H)(\overset{H\,\oplus}{O}CH_3)(OCH_3) \rightleftharpoons CH_3-C(\overset{\oplus}{H})(OCH_3) + HOCH_3 \;\underset{}{\overset{H_2O}{\rightleftharpoons}}\; CH_3-C(H)(\overset{\oplus}{O}H_2)(OCH_3) \longrightarrow$$

Plots of log k vs. H_0 or log C_{H_3O} are common when a quantitative model is being tested. When a more qualitative investigation of pH dependence is appropriate one will sometimes find plots of k vs. pH. A now-classic example comes from the work of Jencks and coworkers on the additon of hydroxylamine to acetone. The striking feature of the k vs. pH plot is the maximum that it exhibits at about pH 4.5 (see Figure 6.2).

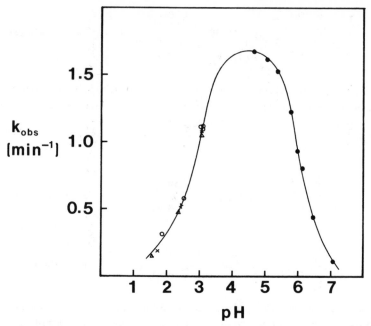

FIGURE 6.2. Dependence of rate on pH for the reaction of acetone with hydroxylamine. Reprinted with permission of W. P. Jencks, *J. Am. Chem. Soc.*, **81**, 475 (1959). Copyright 1959, American Chemical Society.

The explanation for this behavior is, again, a change in rate-determining step with changing pH. A plausible mechanism for the addition reaction is

At pH $<$ 4.5 the hydroxylamine is almost completely protonated and so the addition step is made rate determining by virtue of the limited amount of available nucleophile. At pH $>$ 4.5 the hydroxylamine is mostly unprotonated and so the addition step becomes fast but the dehydration step now suffers from a lack of available protons, causing it to become rate determining.

6.3. GENERAL AND SPECIFIC CATALYSIS BY ACIDS AND BASES

The control of pH for the experiments described in Sections 6.1 and 6.2 is usually achieved by use of a buffer, typically a metal salt MA and the corresponding weak acid HA. In dilute solution where concentrations can be used in place of activities one can write

$$\frac{[H_3O^+][A^-]}{[HA]} = K_a$$

For weak acid

$$[HA] \simeq C_{HA}, \qquad [A^-] \simeq C_{MA}$$

Hence,

$$pH \simeq pK_a + \log(C_{MA}/C_{HA})$$

Studies of acid-catalyzed reactions in which pH was controlled by this device have revealed an interesting dichotomy of behaviors. For many reactions, such as the hydrolysis of acetaldehyde dimethyl acetal, the rate is a function of pH only, at a given temperature. For others, such as the hydrolysis of ethyl vinyl ether, the observed rate constant is a linear function of buffer concentration even at constant pH (i.e., a linear function of C_{MA} at constant C_{MA}/C_{HA}). This difference in behavior can be traced to a difference in rate-determining step of the hydrolysis mechanism. In order to make the connection between mechanism and dependence of the observed rate constant on buffer concentration let us consider the generalized reaction of some substrate S with H_3O^+ and HA:

$$S + H_3O^+ \underset{k_{-1}}{\overset{k_1}{\rightleftharpoons}} SH^+ + H_2O$$

$$S + HA \underset{k_{-2}}{\overset{k_2}{\rightleftharpoons}} SH^+ + A^-$$

$$SH^+ \overset{k_3}{\longrightarrow} Products$$

Applying the steady-state approximation to SH^+, one can write

$$\frac{-d[S]}{dt} = \frac{k_3(k_1[H_3O^+] + k_2[HA])}{k_{-1}[H_2O] + k_{-2}[A^-] + k_3}[S] \tag{1}$$

If proton transfer is rate determining then $k_3 \gg k_{-1}[H_2O]$, $k_{-2}[A^-]$. Hence,

$$\frac{-d[S]}{dt} = (k_1[H_3O^+] + k_2[HA])[S] \tag{2}$$

Had there been several acids, HA_i, the rate law would have been

$$\frac{-d[S]}{dt} = (k_1[H_3O^+] + \sum_i k_{ci}[HA_i])[S]$$

and so this behavior is called *general acid catalysis*. Remembering that

$$[HA] = \frac{C_{MA}}{K_a}[H_3O^+]$$

We can substitute for $[HA]$ in Eq. (2):

$$\frac{-d[S]}{dt} = \left(k_1 \frac{k_2}{K_a} \cdot C_{MA}\right)[H_3O^+][S]$$

Thus at constant pH the observed rate constant will show a linear dependence on C_{MA}.

If we now return to Eq. (1) and consider the situation in which decomposition of SH^+ is rate determining, that is, $k_3 \ll k_{-1}[H_2O]$, $k_{-2}[A^-]$ then

$$\frac{-d[S]}{dt} = \frac{(k_1 + k_2[A^-]/K_a)}{(k_{-1}[H_2O] + k_{-2}[A^-])} \cdot k_3[H_3O^+][S]$$

Since

$$\frac{k_{-1}k_2}{k_1k_{-2}} = \frac{K_a}{[H_2O]}$$

Then

$$\frac{-d[S]}{dt} = \frac{k_1k_3}{k_{-1}[H_2O]} \cdot [H_3O^+][S]$$

Now the rate law shows dependence on $[H_3O^+]$ as the only acid. This behavior is therefore called *specific acid catalysis*. The absence of $[HA]$ in the rate expression means that the observed rate constant will be independent of buffer concentration at constant pH.

In summary, then, general acid catalysis occurs when proton transfer is rate determining. It is detected by a linear dependence of k_{obs} on C_{MA} at constant pH. Specific acid catalysis occurs when proton transfer takes place in a rapid preequilibrium. It is detected by a lack of dependence of k_{obs} on C_{MA} at constant pH.

As one might anticipate, base-catalyzed reactions show a similar dichotomy

General base catalysis: $\quad \dfrac{-d[SH]}{dt} = \left(k_1[HO^-] + + \sum_i k_{ci}[B_i]\right)[SH]$

Specific base catalysis: $\quad \dfrac{-d[SH]}{dt} = k_{obs}[HO^-][SH]$

In acid-catalyzed reactions where a proton is transferred to a heteroatom such as O, N, or S one usually observes specific acid catalysis. The reason is that proton transfers of this kind are usually very rapid and often diffusion controlled, as shown by the work of Eigen in the 1960s.[12] An exception is the acid-catalyzed hydrolysis of orthoacetates where the proton transfer apparently occurs in concert with C—O cleavage.

Proton additions to neutral carbon bases such as olefins apparently require greater structural change in the substrate and are usually rate determining, resulting in general acid catalysis. Again, some exceptions are known.[13]

As always, no experimental technique can *prove* some mechanistic hypothesis to be correct. In the present context it is important to note that observation of a linear dependence of k_{obs} on C_{MA} is a necessary but *insufficient* criterion for general acid catalysis. For many reactions one can construct plausible kinetic schemes involving specific acid catalysis followed by general base catalysis that would show the same behavior. Detailed discussion of this problem and means for resolving it will be discussed in the section dealing with the Brønsted equation (Section 7.4.2).

REFERENCES

1. K. Yates and R. A. McClelland, *Prog. Phys. Org. Chem.*, **11**, 323 (1974).
2. L. P. Hammett and A. J. Deyrup, *J. Am. Chem. Soc.*, **54**, 2721 (1932).

3. W. J. Moore, *Physical Chemistry*, 4th ed., Prentice-Hall, Englewood Cliffs, N.J., 1972, pp. 449–457.

4. J. F. Bunnet and F. P. Olsen, *Can. J. Chem.*, **44**, 1899, 1917 (1966).

5. C. K. Ingold, *Structure and Mechanism in Organic Chemistry*, Cornell University, Ithaca, 1953, p. 754.

6. L. Zucker and L. P. Hammett, *J. Am. Chem. Soc.*, **61**, 2791 (1939).

7. F. A. Long and M. Purchase, *J. Am. Chem. Soc.*, **72**, 3267 (1950).

8. D. McIntyre and F. A. Long, *J. Am. Chem. Soc.*, **76**, 3240 (1954).

9. J. G. Pritchard and F. A. Long, *J. Am. Chem. Soc.*, **78**, 2667 (1956).

10. M. Duboix and A. de Sousa, *Helv. Chim. Acta*, **23**, 1381 (1940).

11. C. H. Rochester, *Acidity Functions*, Academic, London, 1970.

12. M. Eigen, *Angew. Chem. Int. Ed. Engl.*, **3**, 1 (1964).

13. H. Wautier, S. Desauvage, and L. Hevesi, *J. Chem. Soc. Chem. Commun.*, 738 (1981).

CHAPTER 7

INTERPRETATION OF ACTIVATION PARAMETERS

There are three commonly applied equations that purport to describe the temperature dependence of rate constants:

Arrhenius equation:

$$k = Ae^{-E_a/RT}$$

Collision theory equation:

$$k = pZe^{-E^*/RT}$$

$$Z = d_{AB}^2 \sqrt{\frac{8\pi \mathbf{k} T}{m_A m_B}(m_A + m_B)}$$

Transition state theory equation:

$$k = \frac{\mathbf{k} T}{h}e^{\Delta S^{\ddagger}/R}e^{-\Delta H^{\ddagger}/RT}$$

E_a, E^*, and $\Delta H^{\ddagger}$ are usually taken to be temperature independent, in which case all three equations can be summarized in one:

$$k = CT^n e^{-U/RT}$$

Unfortunately, the three models cannot agree on the correct value for n. In the Arrhenius equation $n = 0$, in collision theory $n = 1/2$, and in transition state theory $n = 1$. As it turns out it is the exponential term that really dominates the temperature dependence of k and so any of the three equations can be used to fit a set of experimental data. Nevertheless this inconsistency does serve as a reminder that the choice of activation parameters used to describe a reaction is quite arbitrary. In this chapter we will use the transition state model (Eyring equation) because the parameters $\Delta S^{\ddagger}$ and $\Delta H^{\ddagger}$ can conveniently be related to their thermodynamic analogs ΔS° and ΔH°, making the task of physical interpretation somewhat easier.

7.1. $\Delta H^{\ddagger}$

The enthalpy of activation for a reaction can be determined from the gradient of a plot of $\ln(k/T)$ vs. $1/T$.

$$\ln\left(\frac{k}{T}\right) = \frac{\Delta S^{\ddagger}}{R} + \ln\left(\frac{k}{h}\right) - \frac{\Delta H^{\ddagger}}{RT}$$

If the determination of $\Delta H^{\ddagger}$ for some reaction is to be useful in evaluation of mechanistic hypotheses, one must usually have available some model that allows prediction of the activation enthalpies expected for the various mechanisms under consideration. Sometimes this model takes the form of a sophisticated quantum mechanical calculation but the application of such methods is really outside the scope of this book. Instead we will focus on two much simpler and more empirical models.

7.1.1. Benson Group Additivities

A completely reliable way of estimating heats of formation of molecules would be of enormous value for the study of reaction mechanisms. One could, for example, rigorously exclude a given molecule as a reaction intermediate if its calculated enthalpy of formation exceeded the sum of the activation enthalpy for the reaction and the enthalpy of formation of the reactant(s).

Needless to say there is not, in fact, a method of calculating enthalpies of formation that is completely reliable. Nevertheless, there do exist schemes that allow quite accurate estimates of ΔH_f° for a variety of organic molecules. One method known as Benson group additivity will be developed in this section and will then be applied to some mechanistic problems.

A plausible way of estimating ΔH_f° for the alkanes might be to assume that all C—H and C—C bonds make constant contributions. This hypothesis is easily tested by comparison of calculated and observed ΔH_f° for the first few alkanes.

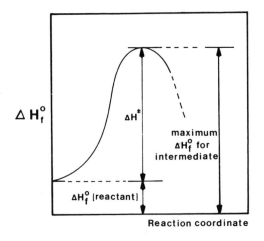

$$\Delta H_f^{\circ}(CH_4) = -17.9 \text{ kcal/mol (experimental)}$$

Assign $\Delta H_f^{\circ}(C-H) = -17.9/4$

$$= -4.5 \text{ kcal/mol}$$

$$\Delta H_f^{\circ}(C_2H_6) = -20.2 \text{ kcal/mol (experimental)}$$

Assign $\Delta H_f^{\circ}(C-C) = -20.2 - 6(-4.5)$

$$= +6.8 \text{ kcal/mol}$$

Now one is in a position to calculate ΔH_f° for the remaining saturated alkanes, for example:

$$\Delta H_f^{\circ}(C_3H_8) = 2\Delta H_f^{\circ}(C-C) + 8\Delta H_f^{\circ}(C-H)$$
$$= -22.4 \text{ kcal/mol (calculated)}$$

A few results can be summarized as follows:

Alkane	ΔH_f° (calc.)	ΔH_f° (obs.)
C_3H_8	-22.4	-24.8
$n\text{-}C_4H_{10}$	-24.6	-30.2
$n\text{-}C_5H_{12}$	-26.8	-35.2
$c\text{-}C_6H_{12}$	-13.2	-29.3

The outcome is not very satisfactory. Although the ΔH_f° calculated for propane is not bad, the discrepancy increases with increasing chain length and the problem is exacerbated by ring closure.

The major flaw in this simple idea is that two-atom information is not enough; not all C—H bonds are the same, for example. One could hope to improve the situation by increasing the number of atoms in the subunits. Benson and coworkers,[1] following this line, define a *group* as an atom with connectivity >1 together with its attached ligands. Thus ethane has two identical groups:

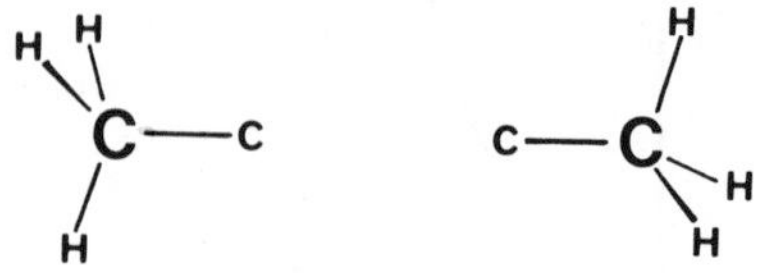

In Benson's notation each group would be written $[C—(C)(H)_3]$. The first letter defines the atom at the center of the group and the letters in parentheses define the ligand atoms.

We can now repeat our previous procedure to develop values for the group contributions and then recalculate ΔH_f^o for some of the higher alkanes.

$$\Delta H_f^o(C_2H_6) = -20.2 \text{ kcal/mol (experimental)}$$

$$\text{Assign} \quad [C—(C)(H)_3] = -10.1 \text{ kcal/mol}$$

$$\Delta H_f^o(C_3H_8) = -24.8 \text{ kcal/mol (experimental)}$$

$$\text{Assign} \quad [C—(C)_2(H)_2] = -24.8 - 2(-10.1)$$

$$= -4.6 \text{ kcal/mol}$$

The enthalpies of formation for the higher n- and c-alkanes can now be calculated (the branched alkanes would require a few more groups).

Alkane	ΔH_f^o (calc.)	ΔH_f^o (obs.)
$n\text{-}C_4H_{10}$	-29.4	-30.2
$n\text{-}C_5H_{12}$	-34.0	-35.2
$c\text{-}C_6H_{12}$	-27.6	-29.3
$c\text{-}C_3H_6$	-13.8	$+12.7$

Clearly, the use of groups has led to a considerable improvement in the calculated ΔH_f^o values over those from the bond additivity approach. The precision in the data for acyclic alkanes can be improved to ± 1 kcal/mol by using best fit group values instead of the values obtained from ethane and propane alone.

The discrepancy in ΔH_f^o calculated for cyclopropane can be attributed to ring strain, and so it becomes necessary to tabulate a list of ring corrections in addition to the group contributions. Benson and coworkers have extended this approach to include heteroatoms and unsaturated compounds. The group contributions and ring corrections are listed in Appendix 5.

Perhaps most useful from the point of view of mechanistic studies is the extension of the group additivity method to include free radicals. It might appear that this would open the way to calculating ΔH_f° for biradicals in thermal rearrangements but there are problems, perhaps best illustrated by the question of cyclopropane stereomutation that we discussed in Chapter 4.

The enthalpy of formation of cyclopropane is known to be 12.7 kca/mol, as we have already seen. The Benson method allows one to calculate the enthalpy of formation of the trimethylene biradical that is implicated in the stereomutation process. The result is ΔH_f° (trimethylene) = 67.2 kcal/mol. Thus the group additivity method suggests that there should be a 54.5 kcal/mol difference in ΔH_f° between trimethylene and cyclopropane. Experimentally the activation enthalpy for stereomutation of cyclopropane-1,2-d_2 is found to be between 59.8 kcal/mol (reference 2) and 63.7 kcal/mol (reference 3), which would seem to imply that the biradical must lie in a potential energy well that is 5.3–9.2 kcal/mol deep.

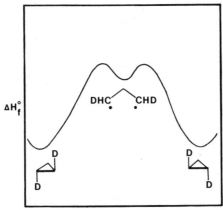

Reaction Coordinate

Were this to be true the biradical would almost certainly execute many rotations about the C—C bonds before reclosing to cyclopropane. It was this kind of analysis that led Benson to propose the random rotation mechanism for stereomutation that we discussed in Chapter 4. Unfortunately, as we saw, the experimental facts are not in accord with this mechanism. So what is wrong?

There seem to be two plausible contributors. The first is that the Benson group equivalents were developed from experiments involving generation of mono-radicals (doublet state molecules). By applying the values to biradicals one is implicitly assuming that there is no interaction between the unpaired electrons. In reality this is not correct, the degenerate doublet states of the unpaired electrons split by interaction into a triplet state and a singlet state, usually with the former lying lower in energy, but by unknown (and presumably variable) amounts. The biradical calculated by using Benson group equivalents is some mythical hybrid of triplet and singlet states. But in a thermal rearrangement one is usually (not always) restricted to the singlet manifold and so it would be the enthalpy of formation for the pure singlet biradical that ought to be used in the mechanistic analysis. The Benson group additivites can thus be expected to have a systematic error that makes singlet biradicals appear to be too stable, unfortunately by an unknown amount.

The second possible problem has recently been identified by Doering.[4] He points out that the enthalpies of formation of the alkyl radicals, on which the group equivalents are based, have traditionally been difficult to measure and have undergone a systematic upward revision with the passage of time. Thus the ΔH_f° values for ethyl, isopropyl, and *tert*-butyl radicals have increased by 2.6, 2.7, and 2.5 kcal/mol, respectively, between 1973, when the set of group equivalents in Appendix 5 was tabulated, and 1981. These apparently small differences can have a rather large effect on the mechanistic interpretation of reactions for which biradicals are implicated. In the case of cyclopropane, an upward revision of ΔH_f° for trimethylene by 5.2 kcal/mol reduces the depth of the potential energy well to 0.1–4.0 kcal/mol. Experimental uncertainty and the spin state problem mentioned above could easily make it disappear altogether.

The conclusions to be drawn from group additivity calculations are clearest when the hypothetical biradical is calculated to have an enthalpy of formation that is substantially greater than that determined for the transition state of a reaction. An example is the retro-Diels–Alder reaction of cyclohexene to butadiene and ethylene. As we have discussed before (Chapters 3 and 5) one could envision a biradical intermediate in this transformation:

The enthalpy of formation calculated for the biradical (including the Doering modification) is 69.0 kcal/mol. Given the experimental ΔH_f° of −0.8 kcal/mol for cyclohexene and the experimental activation enthalpy of 63.9 kcal/mol (reference 5) for the reaction, one can see that the Benson biradical is calculated to be some 6 kcal/mol above the transition state. The true, singlet biradical would presumably be higher still, and so one can reasonably exclude the biradical as an allowable intermediate in the reaction. It is important to emphasize, however, that exclusion by calculation does not carry the same rigor that exclusion by experiment would, simply because of the host of additional assumptions that go into the calculational method.

7.1.2. Substituent Effects on $\Delta H^{\ddagger}$ for Pericyclic and Biradical Reactions

There are certain types of reaction that are difficult to incorporate into the Benson group additivity scheme. One such is the class of pericyclic reactions, including cycloadditions, electrocyclic ring openings, and sigmatropic migrations.

In the retro-Diels–Alder reaction that we discussed at the end of Section 7.1.1, it was possible to exclude the biradical mechanism but it was not possible to demonstrate that the pericyclic alternative was consistent. The problem with this example and with all other pericyclic reactions is that the transition states have no obvious ground state analogs that will allow one to evaluate the group contributions. In this section we will consider an alternative approach that goes some way to resolving the problem.

In 1975 Evans and Golob[6] reported that the Cope rearrangement of compound **1** could be accelerated by the remarkable factor of 10^{17} if the hydroxyl group were simply deprotonated by KH in 18-crown-6. The failure of epimer **2** to rearrange under the same conditions suggested that the mechanism was still that of a concerted [3,3] sigmatropic shift. Later stereochemical studies[7] reinforced this conclusion.

What, then, is the source of this remarkable rate enhancement? A possible answer is provided by the following Born–Haber cycle, which is somewhat unusual in that it includes two transition states:

$$\theta = 2.303\,RT$$

The difference in free energy of activation for the Cope rearrangements of the alcohol and the alkoxide can be calculated from Evans' data. It turns out to be

16.1 kcal/mol at 25°C. Given that

$$\Delta G^{\ddagger}(\text{OH}) - \Delta G^{\ddagger}(\text{O}^-) + 2.303 RT(\text{p}K_a^{\ddagger} - \text{p}K_a) = 0$$

one can deduce the difference in $\text{p}K_a$ between the reactant and transition state of the Cope rearrangement. The difference turns out to be a remarkable 11.8 $\text{p}K_a$ units at 25°C. This is comparable to the difference in acidity between a secondary alcohol and a phenol. Assuming that the reactant alcohol has an acidity comparable to that of most secondary alcohols, one deduces that an alkoxide anion is stabilized to about the same extent in the Cope rearrangement transition state and in a phenol. It is striking that this similarity in stability is paralleled by a similarity in structure. While it is true that the Cope rearrangement transition state is nonplanar, it does possess a cycle of six "π" electrons that is similar to the cyclic array in benzene.[8] One is led to wonder, then, whether the mechanism of stabilization of the alkoxide ion in the Cope rearrangement transition state could be the same as that in a phenol, namely, π-electron delocalization.

This hypothesis is subject to test. The analogy between the transition state of the alkoxy-Cope rearrangement and a phenoxide ion should, according to this model, find a parallel in the cationic domain. The delocalization energies of a phenoxide ion and a benzyl cation are comparable according to most molecular orbital models and so one might expect that the Cope rearrangement would be accelerated just as well by a carbocation in place of the alkoxide ion.

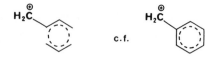

The hypothesis receives support from the work of Breslow and Hoffman[9] who showed that the solvolysis of tosylate **3** proceeds with rearrangement to give the alcohol **4**. If the tosylate leaving group was replaced by iodine and the solvolysis assisted by Ag^+ ion, the reaction could be made to occur rapidly at $-15°$C, probably corresponding to $\Delta G^{\ddagger} \simeq 18$ kcal/mol. This represents an upper limit on the activation free energy of the Cope rearrangement since the rate-determining step is probably the heterolysis of the C—I bond. Given that the $\Delta G^{\ddagger}$ for rearrangement of **5** is 30 kcal/mol at $-15°$,[10] one deduces that the cationic substituent has reduced the $\Delta G^{\ddagger}$ for rearrangement by at least 12 kcal/mol, which is, indeed, comparable to the alkoxide substituent effect.

The verification of the analogy between anionic and cationic substituent effects gives one some confidence in the model. We will now proceed to use it in a more quantitative analysis of structural effects on $\Delta H^{\ddagger}$ for hydrocarbon pericyclic reactions.

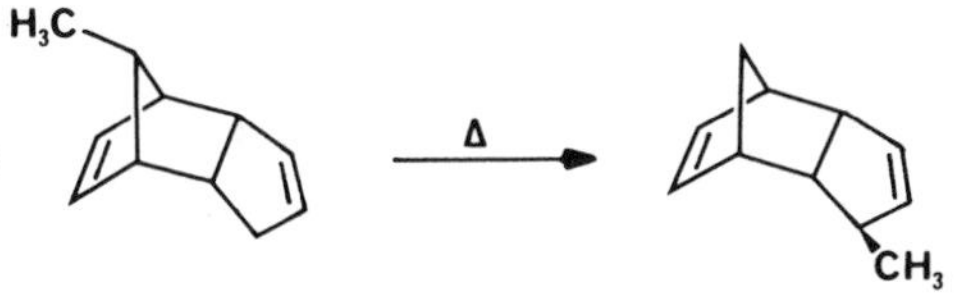

The concept of a "thermodynamic driving force" is one that many organic chemists use explicitly or implicitly. Thus when considering the activation enthalpies for ring opening of cyclobutene and benzocyclobutene, one might be inclined to expect a larger $\Delta H^{\ddagger}$ for the latter because of the loss of aromaticity that results from the bond cleavage. This expectation is borne out in the experimental values:[11,12]

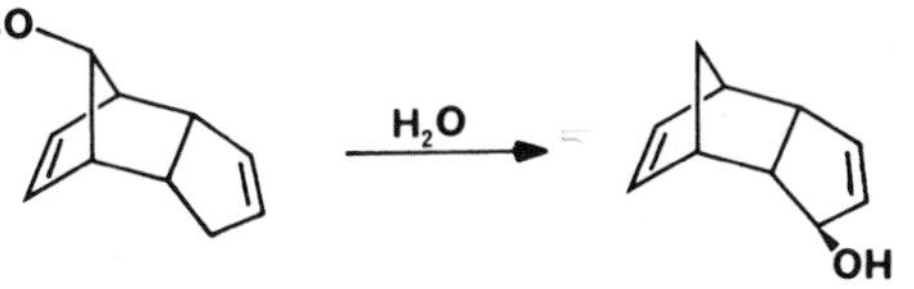

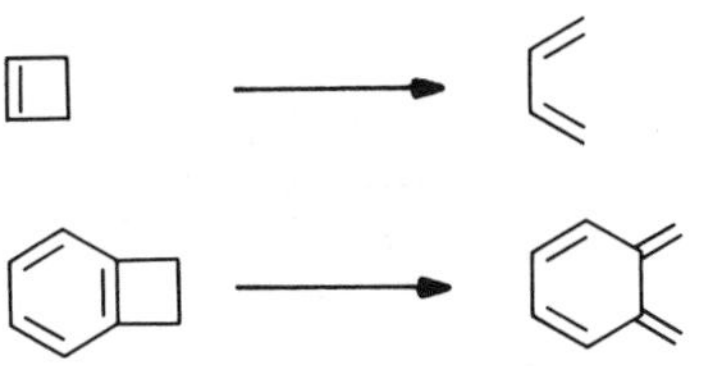

$\Delta H^{\ddagger} \left(kcal/mol \right)$

31.6

38.9

It might, then, be something of a surprise to discover that the relative magnitudes of $\Delta H^{\ddagger}$ are reversed for the ring openings of two apparently similar molecules.[13,14]

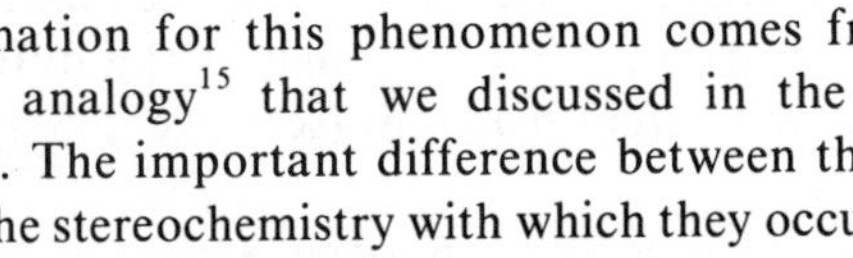

$\Delta H^{\ddagger} \left(kcal/mol \right)$

26.2

18.4

An explanation for this phenomenon comes from application of the π-isoconjugate analogy[15] that we discussed in the context of the Cope rearrangement. The important difference between the two sets of ring opening reactions is the stereochemistry with which they occur. In the unconstrained set, cyclobutene and benzocyclobutene, the Woodward–Hoffmann rules dictate a conrotatory mode.[16] The transition state model for the conrotatory ring opening of cyclobutene is Möbius cyclobutadiene[17] while for benzocyclobutene ring opening it is Möbius benzocyclobutadiene. One can assess the impact of the benzannelation by performing simple Hückel molecular orbital (HMO) calculations on the π electron arrays of reactant and transition state model. If ΔE_{π} is

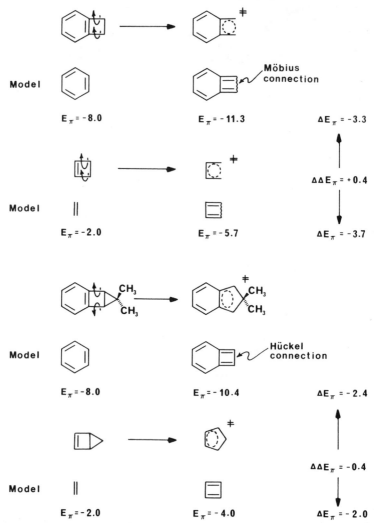

Figure 7.1. Model for analyzing the ring openings of cyclobutene, benzocyclobutene, and their bicyclic analogs. See text for discussion of the calculations.

defined as the difference in HMO π-electron energy between the transition state model and the reactant then $\Delta\Delta E_\pi$ can be defined as the difference in ΔE_π between the benzannelated and parent reactions (see Figure 7.1). The units for $\Delta\Delta E_\pi$ are $|\beta|$.

The fact that $\Delta\Delta E_\pi$ turns out to be positive means that one expects benzannelation to increase $\Delta H^{\ddagger}$ for the ring opening reaction, as observed experimentally.

For the bicyclo[2.1.0]pentene and benzobicyclo[2.1.0]pentene ring openings the conrotatory mode would lead to a product that had a *trans* double bond in a five membered ring. The strain associated with such a structure renders it inaccessible and the reaction follows the "forbidden" disrotatory course instead. The transition state for disrotatory cyclobutene ring opening is modeled by normal (Hückel) cyclobutadiene instead of Möbius cyclobutadiene. This change has a profound effect on the value of $\Delta\Delta E_{\pi}$, which now turns out to be negative, in accord with the observed reduction in $\Delta H^{\ddagger}$ upon benzannelation.[18]

Closer inspection of the calculations described above reveals a striking phenomenon: the ratio of $\Delta\Delta H^{\ddagger}$ (the difference in activation enthalpy for ring opening between the benzannelated and parent molecules) to $\Delta\Delta E_{\pi}$ is virtually identical for the two reactions. For the first it is $(38.9 - 31.6)/0.362 = 20.2$ kcal/(mol$|\beta|$), while the second is $(18.4 - 26.2)/(-0.381) = 20.5$ kcal/(mol$|\beta|$). This observation naturally leads one to wonder whether there might be a correlation between $\Delta\Delta E_{\pi}$ and $\Delta\Delta H^{\ddagger}$ for pericyclic reactions of hydrocarbons. Wilcox and Carpenter investigated this question and found that indeed there was.[19] The correlation was found to include not only pericyclic reactions but also biradical reactions, such as cyclopropane stereomutation, and simple C—C bond homolyses to give discrete free radicals. The plot of $\Delta\Delta H^{\ddagger}$ vs. $\Delta\Delta E_{\pi}$ is shown in Figure 7.2. The line comes from a weighted least-squares analysis in which the dissociations to free radicals were given lower weight because of the greater uncertaintly in their $\Delta H^{\ddagger}$ values. The correlation is described by the equations

$$\Delta\Delta H^{\ddagger} = 20.35\Delta\Delta E_{\pi} \pm 1.65 \text{ kcal/mol (pericyclic and biradical reactions)}$$

$$\Delta\Delta H^{\ddagger} = 20.35\Delta\Delta E_{\pi} \pm 4.37 \text{ kcal/mol (homolysis to free radicals)}$$

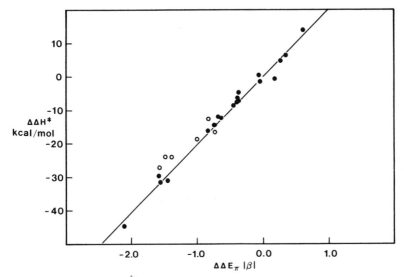

FIGURE 7.2. Graph of $\Delta\Delta H^{\ddagger}$ vs. $\Delta\Delta E_{\pi}$. The open circles represent homolyses to free radicals; the filled circles are pericyclic and biradical reactions. See C. F. Wilcox, Jr. and B. K. Carpenter, *J. Am. Chem. Soc.*, **101**, 3897 (1979) for the reactions used to create this graph.

The existence of this quantitative correlation is somewhat puzzling when one considers the crude nature of the model. Nevertheless, given that it does exist, one can use it for mechanistic studies.

The Cope rearrangement provides an interesting example of how one can use the correlation to detect changes in mechanism. Using a pericyclic transition state model (i.e., benzene) for the [3,3] sigmatropic shift of 1,5-hexadiene, one predicts an increase of 1.6 kcal/mol in $\Delta H^{\ddagger}$ upon replacement of the hydrogens on C2 and C5 with phenyl groups. This prediction is in excellent agreement with the observed[20] $\Delta \Delta G^{\ddagger}$ of 1.5 kcal/mol between the rearrangements of **5** and **6**. It seems reasonable to assume that the entropies of activation would be very similar for the rearrangements of these two compounds and that $\Delta \Delta G^{\ddagger}$ should therefore be approximately equal to $\Delta \Delta H^{\ddagger}$.

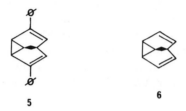

5   6

This close fit to the existing correlation is in strong contrast to the Cope rearrangement of 2,5-diphenyl-1,5-hexadiene. Here the introduction of the phenyl groups causes a *decrease* of 10.2 kcal/mol in $\Delta H^{\ddagger}$.[21] This discrepancy of 11.8 kcal/mol between observed and calculated $\Delta \Delta H^{\ddagger}$ represents more than 7 times the standard deviation in the original correlation and strongly suggests a systematic rather than random error. A plausible explanation is that the mechanism has changed and that for 2,5-diphenyl-1,5-hexadiene the rearrangement proceeds through a cyclohexane-1,4-diyl.[22]

The corresponding biradical might well be inaccessible to **6** because it would entail creating a second cyclopropane ring with its attendant strain energy.

While the explanation presented for the Cope rearrangement is consistent with the experimental observations and the theoretical model outlined in this section, it is perhaps worth reemphasizing that the failure of some new reaction to fit an empirical correlation such as the one used here cannot be taken as rigorous proof of a new mechanism. The Cope rearrangement example is made more convincing by having two sets of experimental observations of which only one set fits the predictions. Had the results on 1,5-hexadiene and its 2,5-diphenyl derivative been the only ones available, the conclusions would have been much less secure.

7.2. $\Delta S^{\ddagger}$

The entropy of activation for a reaction can be determined from the intercept of a plot of $\ln(k/T)$ vs. $1/T$.

$$\ln\left(\frac{k}{T}\right) = \frac{\Delta S^{\ddagger}}{R} + \ln\left(\frac{k}{h}\right) - \frac{\Delta H^{\ddagger}}{RT}$$

For a unimolecular reaction the sign and magnitude of $\Delta S^{\ddagger}$ provide information about the relative degree of order in the reaction transition state and the starting material. Typical values are shown in Figure 7.3.

The homolyses of ethane and di-*tert*-butyl peroxide have positive values for $\Delta S^{\ddagger}$ because of the increase in rotational and translational degrees of freedom that accompany the increase in the number of particles. The negative $\Delta S^{\ddagger}$ observed for ethyl acetate pyrolysis is indicative that something other than simple bond homolysis must be occurring in the rate-determining step. A plausible explanation, consistent with the stereospecific labeling experiment that was discussed in Section 2.2, is that the reaction occurs by way of a cyclic transition state (Figure 7.3). The negative activation entropy then reflects the loss of internal rotational degrees of freedom that must occur in order to allow concerted hydrogen transfer and C—O cleavage. The value of -9 cal mol^{-1} K^{-1} for the activation entropy of HCl elimination from o-methylbenzyl chloride can be attributed to a similar phenomenon.

The conceptually related aliphatic Claisen rearrangement and (Z)-1,3,5-hexatriene ring closure show somewhat different magnitudes for $\Delta S^{\ddagger}$ because in the latter one of the internal rotations is already restricted by the presence of the third double bond.

Benson has assembled a table of group contributions to S° (see Appendix 5) that can be used like the contributions to ΔH_f° discussed in Section 7.1.1. These group contributions can allow one to estimate $\Delta S^{\ddagger}$ for some reactions although, as we saw with $\Delta H^{\ddagger}$ calculations, the values to be assigned to transition state structures are often very uncertain.

For reactions of higher molecularity the sign and magnitude of $\Delta S^{\ddagger}$ depend on the implied standard state,[23] which is defined by the units of the rate constant.

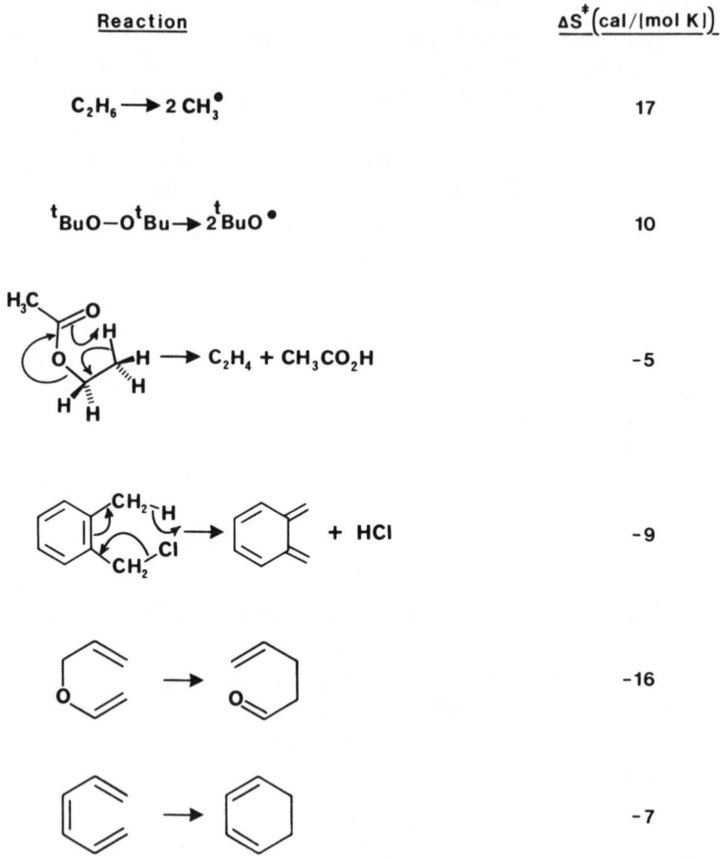

Reaction	$\Delta S^{\ddagger}$ (cal/[mol K])
$C_2H_6 \longrightarrow 2\ CH_3^{\bullet}$	17
$^tBuO-O^tBu \longrightarrow 2\ ^tBuO^{\bullet}$	10

    $\longrightarrow C_2H_4 + CH_3CO_2H$ −5

    $+ HCl$ −9

    −16

    −7

FIGURE 7.3. Typical activation entropies for some unimolecular reactions.

Common standard states for bimolecular reactions are

Units of k	Standard State	$\Delta S^{\ddagger}$ (cal mol^{-1} K^{-1})
L mol^{-1} sec^{-1}	1 mol L^{-1}	X
cm^3 mol^{-1} sec^{-1}	1 mol cm^{-3}	$X + 13.7$
atm^{-1} sec^{-1}	1 atm	$X - 7.9$

Thus it can be seen that neither the sign nor the magnitude of $\Delta S^{\ddagger}$ for a single bimolecular reaction can be of any value in determining mechanisms. If one has a reference bimolecular reaction of "known" mechanism then comparison of its activation entropy to that of some other reaction can be useful (provided they are referred to the same standard state) but under no circumstances does it make sense to compare activation entropies for reactions of different molecularity.

The "structural" interpretation of $\Delta S^{\ddagger}$ for unimolecular reactions breaks down when one encounters the relatively rare circumstance of a surface crossing

7

+ Other dimers

in a thermal reaction. A nice example is provided by the study of Berson and coworkers[24] on the ring opening of 5-isopropylidenebicyclo[2.1.0]pentane (7). The reaction shows first-order kinetics, consistent with a mechanism involving rate-determining formation of the biradical **8**.

8

Dimers

When the temperature dependence of the rate constant was measured, it was found that $\Delta S^{\ddagger} = -16$ cal mol^{-1} K^{-1}. This is an unexpectedly negative value for a reaction involving a bond cleavage. A reasonable comparison is the stereo-isomerization 2-methylbicyclo[2.1.0]pentane[25] for which $\Delta S^{\ddagger} = +2$ cal mol^{-1} K^{-1}.

Earlier work by Berson's group[26] established that **8** had a triplet ground state and so a plausible explanation for the large negative activation entropy might be that the molecule crosses from a singlet to a triplet surface during the ring opening process (see Figure 7.4). The phenomenon of intersystem crossing is one that, for hydrocarbons, requires a mixing of vibrational and electronic states.[27] The number of nuclear configurations that allow such mixing is small. This fact and the small number of "active" vibrations (i.e., those possessing the correct symmetry) within a suitable structure combine to make the intersystem crossing a rather improbable event, as reflected in the negative activation entropy.

7.3. $\Delta V^{\ddagger}$

The use of activation volumes in the study of reaction mechanisms has been the subject of several reviews. Two of the more extensive ones are by Whalley and McCabe and Eckert.[28]

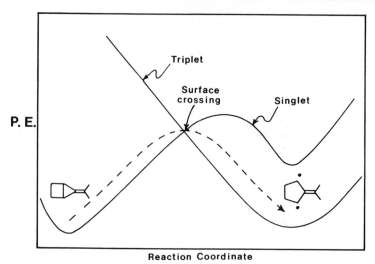

FIGURE 7.4. Surface crossing during the ring opening of 5-isopropylidenebicyclo[2.1.0]pentane as a possible explanation for the observed low activation entropy.

The well-known[29] thermodynamic relationship

$$\left(\frac{\partial \Delta G^\circ}{\partial P}\right)_T = \Delta V^\circ$$

has a close kinetic analogy that can be used to deduce the pressure dependence of a rate constant from the Eyring equation:

$$k = \frac{\mathbf{k} T}{h} e^{-\Delta G^\ddagger / RT}$$

Hence,

$$\ln k = \ln \frac{\mathbf{k} T}{h} - \frac{\Delta G^\ddagger}{RT}$$

$$\left(\frac{\partial \ln k}{\partial P}\right)_T = \frac{-1}{RT} \cdot \left(\frac{\partial \Delta G^\ddagger}{\partial P}\right)_T$$

$$= -\frac{\Delta V^\ddagger}{RT}$$

where $\Delta V^\ddagger$ is the volume of activation. In integral form this becomes

$$\ln k = \ln k^\circ - \frac{\Delta V^\ddagger}{RT} \cdot P$$

Thus one expects that ln k should vary linearly with pressure. This expectation has been verified experimentally in many cases.

Since bonding distances are smaller than van der Waals (nonbonding) distances, the volume occupied by a given assembly of atoms should decrease as the number of bonds between them increases. Thus measurements of $\Delta V^{\ddagger}$ can be expected to give information about the associative or dissociative nature of the transition state for the rate-determining step. In this sense $\Delta V^{\ddagger}$ provides similar information to $\Delta S^{\ddagger}$ except that the latter includes contributions from rotations and vibrations and is thus somewhat more difficult to interpret.

The problem with $\Delta V^{\ddagger}$ measurements that has limited their use so far is one of technical difficulty. Measurements in the gas phase are restricted by the fact that increasing the pressure usually causes a phase transition to the liquid state. Solution phase determination of $\Delta V^{\ddagger}$ is thus the norm, but is made difficult by the very large pressures that are required to see a significant effect—typically in the range of 100—10,000 bar (1 bar = 0.9869 atm = 0.1 MPa). The measurements thus require the use of special high pressure apparatus.[28]

In order to interpret $\Delta V^{\ddagger}$ values it is important to include contributions from the solvent molecules. In solvolysis reactions these might take the form of direct chemical bonding. These effects are relatively easy to anticipate but le Noble[30] has emphasized the contribution from a more insidious phenomenon. When the reaction of a molecule causes some charge separation, the shell of nearest-neighbor solvent molecules usually responds by undergoing a contraction, termed electrostriction, caused by ion–dipole attraction forces. This solvent shell contraction makes a net negative contribution to $\Delta V^{\ddagger}$ even when the reactant itself might be undergoing a dissociation that increases its effective volume.

Typical of the use of $\Delta V^{\ddagger}$ measurements is the study of β-propiolactone hydrolysis in an acidic medium.[31] The experimental value for $\Delta V^{\ddagger}$ is $2.5 \pm 2\,\text{cm}^3$ mol^{-1}. The authors considered a host of different mechanisms, of which just three are shown here:

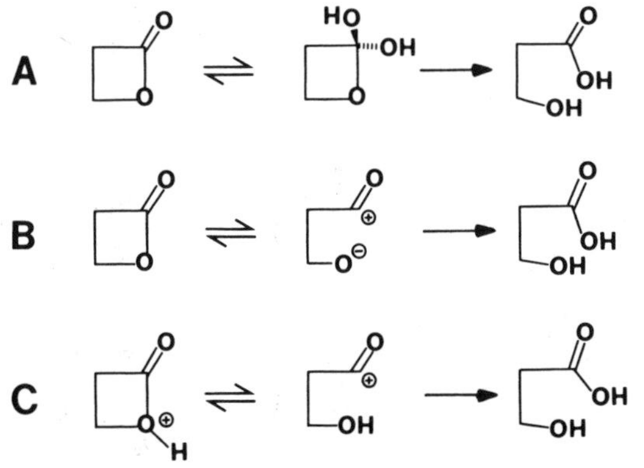

Mechanism A should have $\Delta V^{\ddagger} < 0$ because the transition state for the rate-determining step has incorporated a water molecule. Mechanism B should have $\Delta V^{\ddagger} < 0$ because the charge separation in the transition state should cause electrostriction. Mechanism C remains by exclusion. Note that this is also the mechanism that was favored as a result of the acidity function studies (Section 6.1).

Activation volumes have been determined for a large variety of Diels–Alder reactions.[32] The $\Delta V^{\ddagger}$ values are always large and negative, as one would expect for a bimolecular reaction, but it is somewhat more surprising to find that certain Diels–Alder reactions exhibit activation volumes that are more negative than the overall reaction volumes, ΔV°. Typical values are shown in Table 7.1. This phenomenon has been attributed by some[33] to secondary orbital interactions that bring nominally nonbonded atoms especially close in the transition state. There is little independent evidence to support or refute this hypothesis.

Since the volume of the transition state in a Diels–Alder reaction ought to decrease with increasing number of bonds, one could hope to use $\Delta V^{\ddagger}$ determinations as yet another criterion for the concertedness of the reaction. Concerted formation of both new C—C bonds ought to lead to a more negative $\Delta V^{\ddagger}$ than rate-determining formation of just one bond. The problem, of course, is one of calibration; there is not easy way to determine whether a particular $\Delta V^{\ddagger}$ value is sufficiently negative to indicate concerted formation of two bonds.

An interesting study that presents one possible solution is that of Stewart[34] on chloroprene dimerization. The thermal dimers of chloroprene consist of both cyclohexenes and cyclobutanes. The former could have arisen by a Diels–Alder

TABLE 7.1. Typical Activation and Reaction Volumes for Diels–Alder Reactions

Reaction	$\Delta V^{\ddagger}(cm^3\ mol^{-1})$	$\Delta V^{\circ}(cm^3\ mol^{-1})$
(maleic anhydride) + (CH₃-diene)	-44.7	-33.3
(maleic anhydride) + (Cl-diene)	-41.6	-36.9
(NC-alkene) + (H₃C-diene)	-33.1	-37.0
(CO₂Me-alkyne-CO₂Me) + (cyclopentadiene)	-30.2	-33.9

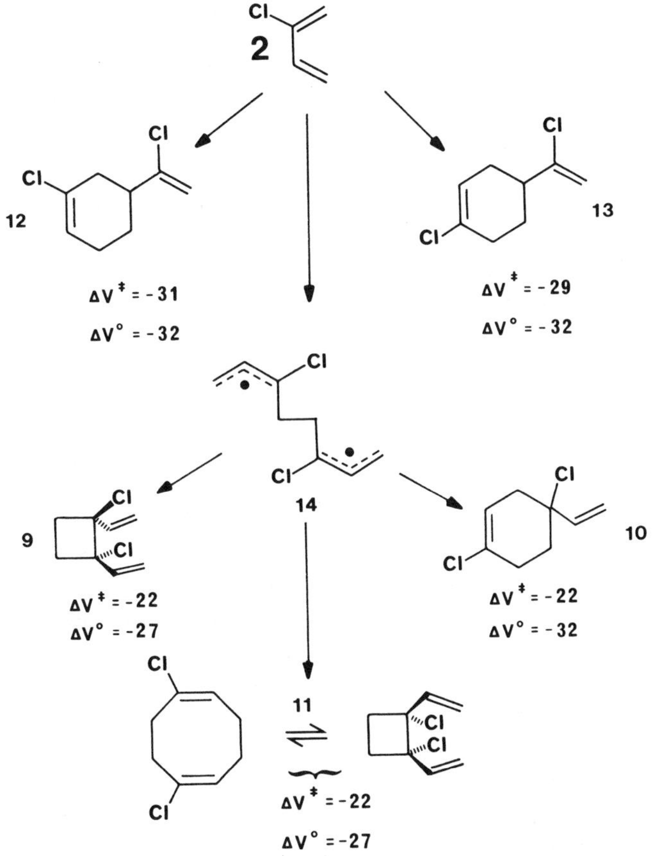

FIGURE 7.5. Activation volumes and a possible mechanism for the dimerization of chloroprene. Units are cubic centimeters per mol for the activation and reaction volumes.

mechanism, the latter might be expected to be formed from a biradical intermediate (see Figure 7.5), if one drew an analogy with the work of Bartlett.[35]

Inspection of the measured activation volumes and the measured or estimated reaction volumes reveals two interesting features. First, products **9–11** are all formed with the same activation volume. This is consistent with (but does not require) a common rate-determining step. Second, the fact that $\Delta V^{\ddagger}$ for these reactions is 5–10 cm^3 mol^{-1} greater than ΔV° is consistent with rate-determining formation of an intermediate in which only one new bond has been made. An obvious candidate is the biradical **14**. By contrast, products **12** and **13** are formed with activation volumes that are comparable to the reaction volumes. This is consistent with a concerted mechanism for the formation of **12** and **13**.

The main concern with this type of study is one of experimental precision. Activation volumes are notoriously difficult to measure and the conclusions derived from them depend on rather small differences. A standard deviation of

± 5 cm^3 mol^{-1} might not be unreasonable, for example, but would render the above conclusions very much less secure.

7.4. $\Delta G^{\ddagger}$—LINEAR FREE ENERGY RELATIONSHIPS

For comparison with some model—say a quantum mechanical calculation—$\Delta G^{\ddagger}$ is the least useful activation parameter since it is both temperature and (in the case of bimolecular reactions) concentration dependent. Nevertheless $\Delta G^{\ddagger}$ is the easiest activation parameter to evaluate since it can be determined from kinetic measurements at a single temperature and pressure. For this reason $\Delta G^{\ddagger}$ has historically been the most frequently measured and interpreted activation parameter in mechanistic studies.

The difficulty of finding models that allow direct interpretation of activation free energies has led to the development of an alternative approach in which changes in $\Delta G^{\ddagger}$ as a result of changes in the medium or the structure of a reactant are compared for two similar reactions. A related approach compares changes in $\Delta G^{\ddagger}$ with changes in ΔG°, either within a single reaction or between different reactions. If the variations in medium or structure are sufficiently small, these studies often result in linear relationships that describe the relative "sensitivities" of the reaction under study and the reference to the particular change. The mechanistic interpretations of these *linear free energy relationships* (LFERs) have been the subject of numerous reviews.[36] In this chapter we will "derive" the general form of a LFER and then consider the application of three specific ones: the Hammett equation, the Brønsted equation and the Grunwald-Winstein equation.

Suppose the overall free energy change for some reaction is ΔG. Variations in ΔG due to alteration of the medium or substitutents, and so on, can be represented as

$$d\Delta G = \left(\frac{\partial \Delta G}{\partial x}\right)dx + \left(\frac{\partial \Delta G}{\partial y}\right)dy + \cdots$$

where $x, y, \ldots$ represents the various parameters that can be changed. If the change in parameter x is small (from x_0 to x_i) and if all the other parameters are kept constant (not so easy in practice!) then one can write

$$\Delta G_i - \Delta G_0 = \left(\frac{\partial \Delta G}{\partial x}\right)(x_i - x_0)$$

or

$$\log\left(\frac{K_i}{K_0}\right) = -\left(\frac{\partial \Delta G}{\partial x}\right)(x_i - x_0)/2.303RT$$

A similar expression can be derived for rate constants rather than equilibrium constants if one makes use of the Eyring equation

$$\Delta G^{\ddagger} = 2.303RT \left(\log \frac{\mathbf{k}T}{h} - \log k \right)$$

$$\Delta G_i^{\ddagger} - \Delta G_0^{\ddagger} = -2.303RT \log\left(\frac{k_i}{k_0}\right)$$

Hence,

$$\log\left(\frac{k_i}{k_0}\right) = -\left(\frac{\partial \Delta G^{\ddagger}}{\partial x}\right)(x_i - x_0)/2.303\,RT$$

The difficulty in trying to use this kind of expression arises when one tries to put units on the parameter x. In what units does one measure the change from a methyl group to a methoxyl group or from ethanol to aqueous acetone? This problem can be sidestepped if one compares effects between two different reactions. To a first approximation one could assume that the change $x_0 \rightarrow x_i$ is the same for two different reactions, A and B. In reality this will never be true but it is reasonable to assume that it will be approximately correct if the two reactions are sufficiently similar. It is then possible to write

$$\log\left(\frac{k_i}{k_0}\right)^{B} = \left\{ \left(\frac{\partial \Delta G^{\ddagger}}{\partial x}\right)^{B} \middle/ \left(\frac{\partial \Delta G^{\ddagger}}{\partial x}\right)^{A} \right\} \log\left(\frac{k_i}{k_0}\right)^{A} \qquad (1)$$

Alternatively, one can compare the kinetic effect on the test reaction B with the thermodynamic effect on reference reaction A:

$$\log\left(\frac{k_i}{k_0}\right)^{B} = \left\{ \left(\frac{\partial \Delta G^{\ddagger}}{\partial x}\right)^{B} \middle/ \left(\frac{\partial \Delta G}{\partial x}\right)^{A} \right\} \log\left(\frac{K_i}{K_0}\right)^{A} \qquad (2)$$

Equations (1) and (2) can be summarized as

$$\log\left(\frac{k_i}{k_0}\right)^{B} = G_i^{AB} X_i$$

where G_i is a parameter describing the relative sensitivities of reactions A and B to the change $x_0 \rightarrow x_i$, and X_i is a parameter describing the effect of this change on the reference reaction, A. The three equations whose applications will constitute the remainder of this chapter can now be seen to be examples of this general form:

Hammett equation (substituent effects)

$$\log\left(\frac{k}{k_0}\right) = \rho\sigma$$

Brønsted equation (acid and base catalysis)

$$\log k_c = \alpha \log K_a + \text{constant}$$

Grunwald-Winstein equation (solvent effects)

$$\log\left(\frac{k}{k_0}\right) = mY$$

We will discuss the application of each of these equations to problems of reaction mechanism in Sections 7.4.1–7.4.3.

7.4.1.   The Hammett Equation

The Hammett equation[36] is an example of a correlation between kinetic data on the test reaction and thermodynamic data on the reference reaction, namely, the dissociation constants of substituted benzoic acids.

$$\sigma = \log\left\{\frac{K_a(XC_6H_4CO_2H)}{K_a(C_6H_5CO_2H)}\right\} \qquad \text{at } 25°C$$

The substituents are restricted to the *meta* or *para* positions since the effects at these sites can be assumed to be purely electronic; *ortho* substituents will have a combination of electronic and steric effects. The parameter σ reflects the relative stabilizing effect of the substituent X on the neutral benzoic acid and its carboxylate anion:

$\sigma > 0$ for substituents more electron withdrawing than hydrogen.

$\sigma < 0$ for substituents less electron withdrawing than hydrogen.

Having established a set of substituent constants (see Table 7.2 for a list) one could now hope that they would apply to other reactions of aromatic compounds and that the graph of $\log(k/k_0)$ vs. σ would be linear with a slope, ρ, that provided a quantitative measure of the sensitivity of the reaction rate to substituent

TABLE 7.2. Hammett Substituent Constants[a]

Substituent	σ_{para}	σ_{meta}	σ^+	σ^-
NH_2	−0.66	−0.16	−1.3	—
OH	−0.37	0.12	−0.92	—
OCH_3	−0.27	0.12	−0.78	−0.2
CH_3	−0.17	−0.07	−0.31	—
$NHCOCH_3$	−0.01	0.21	−0.25	—
C_6H_5	−0.01	0.06	−0.17	—
F	0.06	0.34	−0.07	−0.02
I	0.18	0.35	0.13	—
Cl	0.23	0.37	0.11	—
Br	0.23	0.39	0.15	—
$OCOCH_3$	0.31	0.39	0.18	—
CO_2H	0.45	0.37	0.42	—
CO_2CH_3	0.45	0.37	0.48	0.68
$COCH_3$	0.50	0.38	—	0.87
CF_3	0.54	0.43	0.58	—
CN	0.66	0.56	0.66	0.90
NO_2	0.78	0.71	0.79	1.24
$N(CH_3)_4^+$	0.82	0.88	0.64	—

[a] Values are collected from C. D. Ritchie and W. F. Sager, *Prog. Phys. Org. Chem.*, **2**, 323 (1964) and C. Hansch, A. Leo, S. Unger, K. H. Kim, D. Nakaitani, and E. J. Liem, *J. Med. Chem.*, **16**, 1207 (1973).

change. This phenomenon is, in fact, observed for a wide variety of reactions. The sign and magnitude of ρ allow one to deduce something about charge development at the reaction center:

$\rho > 0$ if the reaction is accelerated by electron withdrawing substituents. This indicates that the aromatic ring has higher electron density in the transition state than in the starting material.

$\rho < 0$ if the reaction is accelerated by electron donating substituents. This indicates that the aromatic ring has lower electron density in the transition state than in the starting material.

Empirically one usually finds $-5 < \rho < 5$ at 25° C. An example is the alkaline hydrolysis of ethyl benzoate

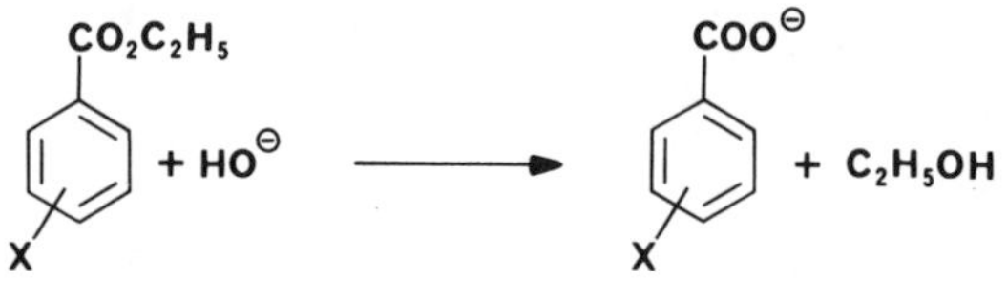

for which $\rho = +2.2$, indicating, as expected, that the ring becomes more negatively charged during the reaction.

   One of the conditions for the existence of a LFER that we discussed in the introduction to this section was that the change $x_0 \rightarrow x_i$ (in this case corresponding to a substituent change) should have qualitatively the same effect on the test reaction and the reference reaction. When this condition is not satisfied the free energy relationship ceases to be linear. Such is the case for substituent effects on reactions where a direct resonance interaction between the substituent and the reaction center is possible, for example,

Since benzoate anions cannot experience a direct resonance interaction with a ring substituent, the normal σ constants cannot be expected to describe such an effect properly. Reactions involving resonance stabilized intermediates of this kind typically show nonlinear Hammett plots.

   In order to circumvent this problem Brown[37] proposed a new set of substituent constants, σ^+ and σ^-, whose definitions allowed for inclusion of resonance effects. The σ^+ parameter was defined from studies on the solvolyses of substituted cumyl chlorides.

$$\sigma^+ = \log\left\{\frac{k(XC_6H_4C(CH_3)_2Cl)}{k(C_6H_5C(CH_3)_2Cl)}\right\} \qquad \text{at } 25°C \text{ in } 90\% \text{ aqueous acetone}$$

   The σ^- parameter was defined from the dissociation constants of substituted phenols.

$$\sigma^- = \log\left\{\frac{K_a(XC_6H_4OH)}{K_a(C_6H_5OH)}\right\} \qquad \text{at } 25°C$$

   It is perhaps worth noting that by choosing σ, σ^+, or σ^- as the substituent parameter in a Hammett plot one is really making an assumption about the mechanism of the reaction. A possible way to avoid this problem is to plot $\log(k/k_0)$ against all three and then to use a statistical criterion, such as a least-squares standard deviation, to determine the best fit.

Swain and Lupton[38] have proposed an alternative approach to that of Brown in which substituent effects are divided into field and resonance effects. By treating these separately one is able to encompass a wider variety of reactions with a single linear free energy relationship but at the expense of an increase in the number of parameters in the equation. A yet finer distinction in which no fewer than four types of resonance effect are considered has also been proposed.[39] Despite these refinements, the original Hammett equation remains the most frequently used. It usually allows one to obtain some qualitative information about electron density changes at the reaction center and often that is all that is required.

When Hammett plots turn out to be curved with a minimum or maximum one can usually be confident that there has been a change in mechanism (or change in rate-determining step) accompanying the change in substituents. A nice example is provided by the work of Rappoport[40] on the reactions of diarylimidoyl chlorides (DIC) with amines. The rate law was found to contain both second-order and third-order terms:

$$\text{Rate} = k'[\text{DIC}][\text{amine}] + k''[\text{DIC}][\text{amine}]^2$$

Recalling the discussion of Section 4.3, we can say that there must be two parallel pathways. One has a rate-determining step with a transition state of composition DIC amine, while the other has a rate-determining step with a transition state of composition DIC (amine)$_2$. Figure 7.6 shows two mechanisms that fit these criteria.

Since the imine nitrogen becomes positively charged in mechanism A but negatively charged in mechanism B one might expect $\rho < 0$ for A and $\rho > 0$ for B. In fact the plots of $\log k'$ and $\log k''$ vs. σ_Y were both curved with minima near $\sigma = 0.4$. A plausible explanation is that mechanism A occurs for substituents at site Y that have $\sigma < 0.4$ (i.e., are more electron donating than Br) and that mechanism B occurs for substituents with $\sigma > 0.4$.

If this explanation were correct one would expect to see a large negative ρ for substituents at site X when mechanism A occurred but a small positive ρ when mechanism B occurred. This is exactly what was observed when piperidine was used as the amine nucleophile.

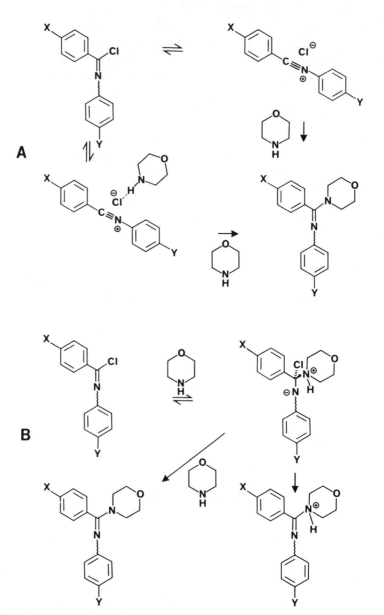

FIGURE 7.6. Two possible mechanisms for the reactions of diarylimidoyl chlorides with morpholine.

Consequences of the Isokinetic Relationship. Within a series of similar reactions one often finds that $\Delta H^\ddagger$ and $\Delta S^\ddagger$ are linearly related:

$$\Delta H^\ddagger = \Delta H_0^\ddagger + \beta \Delta S^\ddagger \tag{3}$$

where β, the slope of the line, has units of temperature and is usually called the *isokinetic temperature*.

This relationship has important consequences for the Hammett equation (and all other LFERs), as can be seen if one recalls the identity

$$\rho = \frac{(\partial \Delta G^\ddagger / \partial \sigma)}{(\partial \Delta G / \partial \sigma)}$$

The denominator can be evaluated from the definition of σ:

$$\frac{\partial \Delta G}{\partial \sigma} = -2.303\,RT\,\frac{\partial \log K_a(XC_6H_4CO_2H)}{\partial \sigma}$$

$$= -2.303\,RT$$

Hence,

$$\rho = \frac{-(\partial \Delta G^\ddagger / \partial \sigma)}{2.303\,RT} \tag{4}$$

$$= \frac{-(\partial \Delta H^\ddagger / \partial \sigma - T\partial \Delta S^\ddagger / \partial \sigma)}{2.303\,RT}$$

If one now differentiates the isokinetic equation [Eq. (3)] with respect to σ:

$$\frac{\partial \Delta H^\ddagger}{\partial \sigma} = \beta\,\frac{\partial \Delta S^\ddagger}{\partial \sigma}$$

Hence,

$$\rho = \frac{(T/\beta - 1)}{2.303\,RT} \cdot \frac{\partial \Delta H^\ddagger}{\partial \sigma}$$

The mechanistically important information is in $\partial \Delta H^\ddagger / \partial \sigma$. Provided that $T < \beta$, ρ and $\partial \Delta H^\ddagger / \partial \sigma$ will bear the same sign relationship that ρ and $\partial \Delta G^\ddagger / \partial \sigma$ do [Eq. (4)]. However, as T, the reaction temperature, approaches β, the isokinetic temperature, $\rho \to 0$ regardless of the sign and magnitude of $\partial \Delta H^\ddagger / \partial \sigma$. At higher temperatures still the sign relationship between ρ and $\partial \Delta H^\ddagger / \partial \sigma$ changes from negative to positive, meaning that the physical interpretation that we have previously put on the sign of ρ should similarly be inverted! With measurements made at a single temperature, as Hammett plots usually are, one can never be absolutely sure of the correct interpretation of ρ.

In fact things are not quite this bad. The weight of considerable empirical evidence[41] suggests that the isokinetic temperatures of most organic reactions are substantially above room temperature. Since most Hammett plots are constructed from rate constants measured at room temperature, the interpretation of ρ that we have discussed remains valid. But problems can arise with Hammett studies conducted at higher temperature. An interesting example is the rearrangement of α-trimethylsilyl benzophenone.

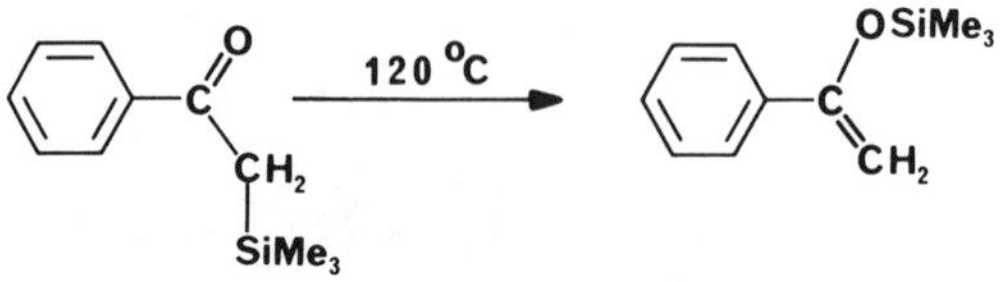

The reaction has been shown to be intramolecular by double-labeling crossover experiments and has also been shown to proceed with retention at silicon.[42] The rearrangement is a formal [1,3] sigmatropic shift, although it is much more facile than the corresponding carbon migration and its stereochemistry is unexpected, requiring an antarafacial migration if it is to be in accord with the Woodward-Hoffmann rules. An alternative mechanism that would explain the data involves formation of a zwitterionic intermediate containing pentacoordinate silicon:

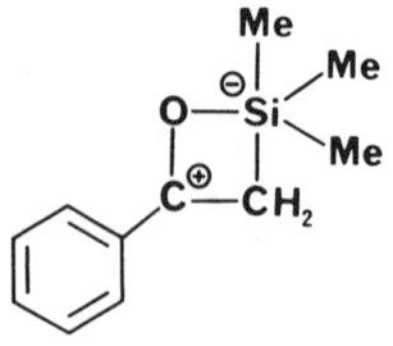

Such an intermediate should be detectable by a large negative ρ in a Hammett plot using substituents on the aromatic ring. When this experiment was carried out it was found that ρ was only -0.78, appearing to disfavor the zwitterion mechanism.[43] However subsequent studies[44] of the temperature dependence allowed determination of β which was found to be $380 \pm 27°$ K, perilously close to the reaction temperature of $393°$ K. The small value of ρ could no longer be considered to invalidate the zwitterion mechanism. In fact solvent effect studies showed a substantial increase in the rate of the reaction in polar media[44] as expected for a reaction involving charge separation in the rate-determining step. Silicon isotope effect studies were also considered to be in accord with the zwitterion mechanism.[44]

7.4.2. The Brønsted Equation

In Chapter 6 we saw that acid-catalyzed reactions fall into two classes—the general acid and specific acid types. Base-catalyzed reactions exhibit a corresponding dichotomy. In the case of general acid catalysis each acid present will catalyze the reaction in question with its own characteristic rate constant, k_c. One might guess that the magnitude of k_c would depend on the strength of the acid, in other words, its equilibrium dissociation constant. Brønsted showed that indeed this was true in many cases, and that k_c and K_a describe a linear free energy relationship.

$$\text{For acids:}\quad \log k_c = \alpha \log K_a + \text{constant}$$

$$\text{For bases:}\quad \log k_c = \beta \log K_b + \text{constant}$$

The parameters α and β are reaction constants analogous to the Hammett ρ, but what is their significance? Empirically one finds that they fall in the range of 0–1 (with a few notable exceptions[45]) and one can gain some insight into their significance by considering the extremes of this range.

When $\alpha = 1$ the structural features of the acid catalyst that determine its strength are fully expressed in the transition state for the proton transfer. (Remember that for general-acid-catalyzed reactions proton transfer is rate determining.) This is most easily explained if the proton transfer is essentially complete in the transition state. By contrast when $\alpha = 0$ the structural features that determine the stength of the acid catalyst have no opportunity for expression in the transition state. This would be consistent with a transition state in which essentially no proton transfer had occurred. It seems a logical interpolation to suggest that intermediate values of α (or β for bases) reflect intermediate degrees of proton transfer or, put another way, that α and β are a measure of the location of the transition state along the reaction coordinate. This suggestion has been frequently espoused[46] and frequently criticized.[47]

Le Noble has pointed out that a two-dimensional potential energy profile (i.e., potential energy as the ordinate and "reaction coordinate" as the abscissa) can be modeled reasonably well by a fourth-order polynomial.[48] If one specifies constant temperature and concentration then the same function could be used to represent the free energy change in a reaction. From such a model one can easily show that the Brønsted α is not identical with the position of the transition state along the reaction coordinate. Rather one finds[49]

$$\alpha = x^{\ddagger 2}(3 - 2x^{\ddagger})$$

where x is defined between 0 and 1 and represents the combination of geometrical parameters that are called the reaction coordinate. $x^{\ddagger}$ is the value of x at which the transition state occurs (i.e., at which the free energy is at a maximum). The interesting thing about this unpromising looking equation is that it approximates $\alpha = x^{\ddagger}$ in the range $0 < x^{\ddagger} < 1$:

$x^{\ddagger}$	"α"
0.00	0.00
0.10	0.03
0.20	0.10
0.30	0.22
0.40	0.35
0.50	0.50
0.60	0.65
0.70	0.78
0.80	0.90
0.90	0.97
1.00	1.00

Miller[50] has extended this kind of investigation to include other functional models and is able to deduce a general equation for the location of the transition state.

In Chapter 6 we noted that the criterion of linear dependence of $\log k_{obs}$ on the buffer concentration could not be taken as proof of a general acid-catalyzed reaction mechanism because similar behavior could be predicted for a reaction involving specific acid catalysis followed by general base catalysis. We will now verify this assertion and then show how the Brønsted equation has been used to resolve the ambiguity in at least one case.

Two plausible mechanisms for the addition of the nucleophile NuH to the carbonyl compound S are shown in Figure 7.7. Mechanism A is genuine general acid catalysis in which a proton is transferred from the acid-catalyst HA to the carbonyl compound in the rate-determining step. Its rate law is

$$-\frac{d[S]}{dt} = k_{obs}[NuH][S]$$

$$k_{obs} = k_1[HA]$$

In mechanism B there is specific acid catalysis followed by rate-determining proton removal in which the anion A^- acts as a general base. Its rate law is

$$-\frac{d[S]}{dt} = k_{obs}[NuH][S]$$

$$k_{obs} = \frac{k_1 k_2}{k_{-1}}[HA]$$

which is experimentally indistinguishable from that for mechanism A.

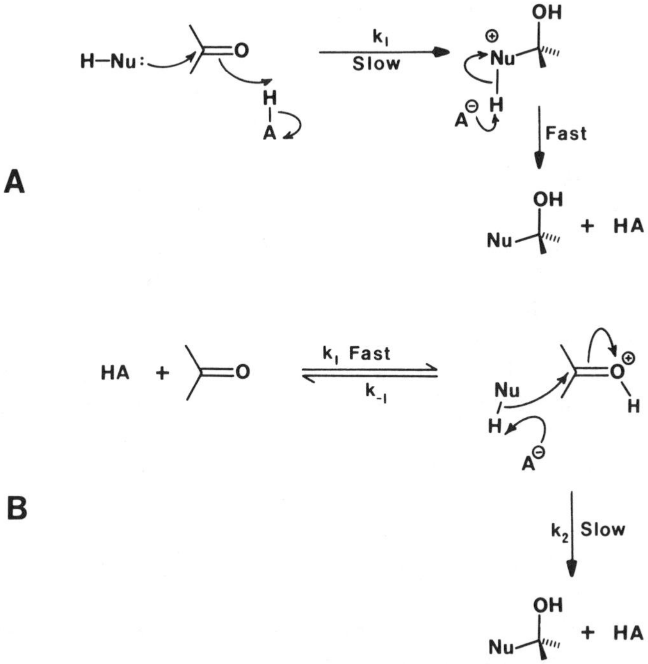

FIGURE 7.7. Two possible mechanism for the addition of a nucleophile to a carbonyl compound.

From the Brønsted catalysis equation one can write

$$k_{obs} = C\{K_a(HA)\}^\alpha$$

for mechanism A. For mechanism B the Brønsted equation is

$$k_2 = C'\{K_b(A^-)\}^\beta$$

But

$$K_b = \frac{K_w}{K_a}$$

where

$$K_w = [H_3O^+][HO^-]$$

Hence,

$$k_2 = C''\{K_a(HA)\}^{-\beta}$$

Since

$$\frac{k_1}{k_{-1}} = \frac{K_a(\text{HA})}{K_a(\text{SH}^+)}$$

$$k_{\text{obs}} = C'''\{K_a(\text{HA})\}^{1-\beta}$$

This looks just like the Brønsted relationship for mechanism A if one makes the substitution $\alpha_{\text{apparent}} = 1-\beta$.

It is possible to distinguish between the two mechanisms, however, by studying the dependence of α_{apparent} on the nucleophilicity of NuH. In mechanism A the transition state should come earlier as NuH becomes more nucleophilic and so α should decrease. In mechanism B, β should decrease with increasing nucleophilicity of NuH but this will cause an *increase* in α_{apparent}. Jencks and coworkers[51] have studied the variation in α_{apparent} with nucleophilicity for addition to a number of different aldehydes. The results show a consistent decrease in α with increasing nucleophilicity, supporting mechanism A over mechanism B.

7.4.3. The Grunwald-Winstein Equation

Solvent–solute interactions are conveniently neglected by most organic chemists in their discussions of reaction mechanisms. One need only look at the effect on a typical heterolytic bond cleavage, however, to recognize that ignoring solvent interactions is hazardous at best.

$$\text{At } 25\,^\circ\text{C} \begin{cases} \Delta H\,(\text{g}) & = 150 \text{ kcal/mol} \\ \Delta H\,(\text{H}_2\text{O}) = & 20 \text{ kcal/mol} \end{cases}$$

Here the effect is presumably a combination of easier charge separation in the polar environment and specific solvation by hydrogen bonding or ion–dipole interaction. At our present stage of understanding there is little hope of being able to quantify such effects on the basis of any nonempirical model. Rather this seems to be another situation where empirical correlations are the likely to be the most fruitful approach.

With this idea in mind Grunwald and Winstein[52] set up a linear free energy relationship whose form is identical to that of the Hammett equation:

$$\log\left(\frac{k}{k_0}\right) = mY$$

where m is the reaction parameter, analogous to the Hammett ρ and Y is the solvent parameter, analogous to σ. *Tert*-butyl chloride solvolysis was used as the standard reaction ($m = 1$) in order to set up the scale of Y values and 80% aqueous ethanol was defined to be the solvent for which $Y = 0$. Hence,

$$Y = \log\left\{\frac{k(\text{test solvent})}{k(80\% \text{ EtOH}/\text{H}_2\text{O})}\right\} \quad \text{at } 25°\text{C}$$

With the Y values for various solvents established (see Table 7.3), one can investigate the solvent dependence of other solvolysis reaction rates. The m values so determined provide some information about the charge separation in the transition state. For example *tert*-butyl bromide solvolysis shows $m = 0.90$, in accord with expectations from the Hammond postulate (the rate-determining step should be less endothermic for *tert*-butyl bromide than for *tert*-butyl chloride and so the transition state should occur earlier, with less charge separation). As one might expect, solvolyses of primary or secondary substrates

TABLE 7.3. Solvent Polarity Parameters

Solvent	Y^a	Z^b	E_T^c
Water	3.49	94.6	63.1
Formamide	0.60	83.2	56.6
Methanol	−1.09	83.6	55.5
Ethanol/water (80%)	[0]	84.8	53.6
Ethanol	−2.03	79.6	51.9
Isopropanol	−2.73	76.3	48.6
Acetonitrile		71.3	46.0
Dimethylsulfoxide		71.1	45.0
Sulfolane		77.5	44.0
Tert-butanol	−3.26	71.3	43.9
Dimethylformamide		68.5	43.8
Acetone		65.7	42.2
Dichloromethane		64.2	41.1
Pyridine		64.0	40.2
Chloroform		63.2	
Hexamethylphosphoramide		62.8	40.9
Tetrahydrofuran			37.4
Diethyl ether			34.6
Benzene		54	34.5
Toluene			33.9
Carbon tetrachloride			32.5
Hexane			30.9

[a] Reference 52.
[b] Reference 53.
[c] Reference 54.

show poor correlations of $\log(k/k_0)$ with Y since in these cases there is presumably a large S_N2 component.

The major drawback to the Grunwald–Winstein scheme is the limited range of its applicability. Since *tert*-butyl chloride solvolyzes at an appreciable rate only in rather polar solvents, no Y values can be obtained for common solvents such as hexane, toluene, or diethyl ether.

In an attempt to devise a solvent polarity parameter with wider ranging application Kosower[53] and Dimroth[54] both turned to ultraviolet spectroscopy. Kosower recognized that when the pyridinium salt **15** absorbed an ultraviolet photon it underwent a charge transfer transition to give the essentially nonpolar radical pair **16**. One could expect that polar solvents would tend to stabilize the ground state with respect to the electronic excited state, thereby causing a hypsochromic shift in the absorption spectrum. Kosower proposed that the energy of the transition in kilocalories per mole [designated Z and measured as $2.859 \times 10^4/(\lambda_{max}$ in nanometers)] could be used as a measure of the solvent polarity.

Dimroth used a similar approach but employed the zwitterion **17** in place of Kosower's pyridinium salt. The improved solubility characteristics of **17** allowed the determination of solvent polarity for media such as toluene and hexane. Kosower's Z values and Dimroth's E_T values are listed in Table 7.3.

As one might expect, there is a good linear correlation between Z and E_T, whereas Z and Y show no obvious correlation. In fact, Z and Y *are* linearly related for a sufficiently narrow range of solvents such as aqueous ethanol mixtures of various proportions,[53] but the breakdown of this relationship over a wider range serves to emphasize that no single model for solvent "polarity" can be expected to serve all of the needs of the mechanistic organic chemist.

REFERENCES

1. S. W. Benson, *Thermochemical Kinetics*, 2nd ed., Wiley-Interscience, New York, 1976.

2. E. V. Waage and B. S. Rabinovitch, *J. Phys. Chem.*, **76**, 1695 (1972).

3. B. S. Rabinovitch, E. W. Schlag, and K. B. Wiberg, *J. Chem. Phys.*, **28**, 504 (1958).

4. W. von E. Doering, *Proc. Natl. Acad. Sci.*, **78**, 5279 (1981).

5. M. Uchiyama, T. Tomioka, and A. Amano, *J. Phys. Chem.*, **68**, 1878 (1964); D. C. Tardy, R. Ireton, and A. S. Gordon, *J. Am. Chem. Soc.*, **101**, 1508 (1979).

6. D. A. Evans and A. M. Golob, *J. Am. Chem. Soc.*, **97**, 4765 (1975).

7. D. A. Evans and J. V. Nelson, *J. Am. Chem. Soc.*, **102**, 774 (1980).

8. B. K. Carpenter, *Tetrahedron*, **34**, 1877 (1978).

9. R. Breslow and J. M. Hoffman, Jr., *J. Am. Chem. Soc.*, **94**, 2111 (1972).

10. V. A. Maronov, I. M. Fadeeva, and A. A. Akhrem, *Izv. Akad. Nauk, SSSR Ser. Khim.*, 436 (1968); 859 (1968).

11. W. R. Roth, H. Bierman, H. Dekker, R. Jochems, C. Mosselman, and H. Hermann, *Chem. Ber.*, **111**, 3892 (1978).

12. W. Cooper and W. D. Walters, *J. Am. Chem. Soc.*, **80**, 4220 (1958).

13. C. F. Wilcox, Jr., B. K. Carpenter, and W. R. Dolbier, Jr., *Tetrahedron,* **35**, 707 (1979).

14. J. I. Brauman and D. M. Golden, *J. Am. Chem. Soc.*, **90**, 1920 (1968).

15. M. J. S. Dewar and R. C. Doughert, *The PMO Theory of Organic Chemistry*, Plenum, New York, 1975, p. 338ff.

16. R. B. Woodward and R. Hoffmann, *The Conservation of Orbital Symmetry*, Academic, New York, 1970.

17. H. E. Zimmerman, *Acc. Chem. Res.*, **4**, 272 (1971).

18. C. F. Wilcox, Jr., B. K. Carpenter, and W. R. Dolbier, Jr., *Tetrahedron,* **35**, 707 (1979).

19. C. F. Wilcox, Jr. and B. K. Carpenter, *J. Am. Chem. Soc.*, **101**, 3897 (1979).

20. F. A. L. Anet and G. E. Schenck, *Tetrahedron Lett.*, 4237 (1970); H. Kessler and W. Ott, *J. Am. Chem. Soc.*, **98**, 5014 (1976).

21. M. J. S. Dewar and L. E. Wade, Jr., *J. Am. Chem. Soc.*, **95**, 290 (1973); **99**, 4417 (1977).

22. W. von E. Doering, V. G. Toscano, and G. H. Beasley, *Tetrahedron,* **27**, 5299 (1971).

23. P. J. Robinson, *J. Chem. Educ.*, **55**, 509 (1978).

24. M. Rule, M. G. Lazzara, and J. A. Berson, *J. Am. Chem. Soc.*, **101**, 7091 (1979).

25. J. P. Chesick, *J. Am. Chem. Soc.*, **84**, 3250 (1962).

26. J. A. Berson, R. J. Bushby, J. M. McBride, and M. Tremelling, *J. Am. Chem. Soc.*, **93**, 1544 (1971).

27. N. J. Turro, *Modern Molecular Photochemistry*, Benjamin/Cummings, Menlo Park, Calif., 1978, p. 185ff.

28. E. Whalley, *Adv. Phys. Org. Chem.*, **2**, 93 (1974); J. R. McCabe and C. A. Eckert, *Acc. Chem. Res.*, **7**, 251 (1974).

29. W. J. Moore, *Physical Chemistry*, 3rd ed., Prentice-Hall, Englewood Cliffs, N.J., 1962, p. 120.

30. W. J. le Noble, *Prog. Phys. Org. Chem.*, **5**, 207 (1967).

31. R. J. Withey, J. E. McAlduff, and E. Whalley, *Proc. Symp. Phys. Chem. High Pressures London*, 1962, p. 196.

32. J. R. McCabe and C. A. Eckert, *Acc. Chem. Res.*, **7**, 251 (1974).

33. A quite detailed discussion of this can be found in reference 32.

34. C. A. Stewart, *J. Am. Chem. Soc.,* **84**, 117 (1962).

35. P. D. Bartlett, *Pure Appl. Chem.*, **27,** 597 (1971).

36. R. D. Topsom, *Prog. Phys. Org. Chem.*, **12,** 1 (1976); S. H. Uger and C. Hansch, *Ibid.*, **12,** 91 (1976); L. S. Levitt and H. F. Widing, *Ibid.*, **12,** 119 (1976).

37. L. M. Stock and H. C. Brown, *Adv. Phys. Org. Chem.*, **1,** 35 (1963).

38. C. G. Swain and E. C. Lupton, *J. Am. Chem. Soc.*, **90,** 4328 (1968).

39. J. Bromilow and R. T. Brownlee, *J. Org. Chem.*, **44,** 1261 (1979) and references therein.

40. R. Ta-Shma and Z. Rappoport, *J. Am. Chem. Soc.*, **99,** 1845 (1977).

41. J. E. Leffler, *J. Org. Chem.*, **20,** 1202 (1955).

42. A. G. Brook, D. M. MacRae, and W. W. Limburg, *J. Am. Chem. Soc.*, **89,** 5493 (1967).

43. A. G. Brook, *J. Organomet. Chem.*, **86,** 185 (1975).

44. G. L. Larson and Y. V. Fernandez, *J. Organomet. Chem.*, **86,** 193 (1975); H. Kwart and W. E. Barnette, *J. Am. Chem. Soc.*, **99,** 614 (1977).

45. F. G. Bordwell, W. J. Boyle, Jr., J. A. Hautala, and K. C. Yee, *J. Am. Chem. Soc.*, **91,** 4002 (1969).

46. See, for example, J. E. Leffler and E. Grunwald, *Rates and Equilibria of Organic Reactions*, Wiley, New York, 1963, p. 235.

47. R. A. Marcus, *J. Am. Chem. Soc.*, **91,** 7224 (1969); A. J. Kresge, *Ibid.*, **92,** 3210 (1970).

48. W. J. le Noble, A. R. Miller, and S. D. Haman, *J. Org. Chem.*, **42,** 338 (1977).

49. B. K. Carpenter, unpublished results.

50. A. R. Miller, *J. Am. Chem. Soc.*, **100,** 1984 (1978).

51. W. P. Jencks, *Catalysis in Chemistry and Enzymology*, McGraw-Hill, New York, 1969, pp. 195–197.

52. E. Grunwald and S. Winstein, *J. Am. Chem. Soc.*, **70,** 846 (1948).

53. E. M. Kosower, *J. Am. Chem. Soc.*, **80,** 3253 (1958).

54. K. Dimroth, C. Reichardt, T. Siepman, and F. Bohlmann, *Ann. Chem.*, **661,** 1 (1963); C. Reichardt, *Angew. Chem. Int. Ed. Engl.*, **4,** 29 (1965).

CHAPTER 8

DIRECT DETECTION
OF REACTIVE INTERMEDIATES

In the methods that were described in Chapters 2–7 the object was to choose among mechanistic hypotheses on the basis of experimental evidence that was often rather indirect. In the cases where an intermediate was postulated to exist we assumed that this intermediate was not directly observable and that its presence or absence had to be inferred by some logically more tortuous procedure. This situation is still the most common in organic chemistry and is the reason that isotopic labeling studies and kinetic measurements are still so important. Nevertheless there do exist conditions under which one can gain more direct information about the existence and properties of transient intermediates. The increasing sophistication of instrumental analysis techniques suggests that this approach to studying reaction mechanisms might become even more important in the future.

8.1. MATRIX ISOLATION INFRARED SPECTROSCOPY

Often molecules that are stable with respect to unimolecular decomposition are still difficult to isolate and characterize because of their high reactivity in bimolecular processes (dimerization or reaction with the solvent, for example). This appears to be the case for many carbenes, carbonium ions, and carbanions as well as for special reactive intermediates such as cyclobutadiene and benzyne. It would seem, then, that one could hope to isolate and observe these species if their bimolecular reactions could be suppressed.

This is the idea behind matrix isolation. The technique involves generation of the reactive intermediate in a solid matrix, often of an inert gas such as argon.

The solid phase suppresses diffusion and slows down dimerization reactions to negligible levels. The matrix itself is chemically inert and transparent to infrared (IR) radiation, allowing one to obtain IR spectra of the substrate.

One difficulty associated with this technique is that rather specialized apparatus is required, a fact that dissuades many investigators from applying it routinely. It is also limited to reactive intermediates that can be prepared by unimolecular photofragmentation since the low temperatures required to maintain a solid matrix obviously preclude the use of any thermal reactions. Finally, the interpretation of the IR spectra in terms of structural properties of the reactive intermediate can be difficult. It is invariably the case that byproducts are produced during the photolysis and trapped in the matrix. These will usually have IR bands of their own and might even interact with the reactive intermediate to form some sort of complex whose structure would not be representative of the isolated molecules. A thorough investigation thus requires the use of several precursors in order to identify which spectral bands belong to the molecule of interest and to rule out the possibility of complex formation.

Perhaps the most thorough investigation of this kind was that on the structure of cyclobutadiene. The early experiments were performed independently and almost simultaneously by the groups of Krantz[1] and Chapman.[2] Both used the bicyclic lactone **1** as their precursor to cyclobutadiene.

The intention was to distinguish between square (D_{4h}) and rectangular (D_{2h}) structures for cyclobutadiene by observing its IR spectrum. In principle this could be done simply by counting the number of bands since group theory allows one to predict that the D_{4h} structure should have four absorption bands whereas the D_{2h} structure should have seven. In reality the C—H stretching vibrations might be expected to be weak and perhaps difficult to observe. Still, one should see three bands below 2000 cm^{-1} for D_{4h} and five for D_{2h}. Both groups reported seeing three bands in this region, Krantz at 573, 653, and 1236 cm^{-1} and Chapman at 570, 650, and 1240 cm^{-1}. The data thus appeared to fit the expectations for a square structure.

The first doubt was cast by Maier[3] who showed that photolysis of the tricyclic precursor **2** gave cyclobutadiene, which exhibited the 570 and 1240 cm^{-1} bands but nothing at 650^{-1}.

He proposed that the 650^{-1} band was really a bending mode of carbon dioxide (the byproduct from photolysis of **1**) in loose association with the cyclobutadiene. This proposal was later confirmed by Krantz[4] who ^{13}C labeled the carbonyl carbon of **1** and showed that the 650 cm^{-1} band exhibited an isotope shift. Thus the real IR spectrum of cyclobutadiene apparently exhibited too few bands for either the square or rectangular structures. In other words the number of bands could not be used to distinguish between the two structures. On the other hand Krantz and Newton[5] had carried out a vibrational analysis that showed that the positions of the observed bands were consistent with the expectations for a square structure using a reasonable set of force constants. A particularly important conclusion from this analysis was the assignment of the 1240 cm^{-1} band to a C—C stretching vibration. Had cyclobutadiene been rectangular one would have expected two C—C stretching vibrations to be IR active. But the assignment of the 1240 cm^{-1} band was challenged by Kollmar and Staemmler[6] and Hess and Schaad[7] who independently carried out *ab initio* molecular orbital calculations on rectangular cyclobutadiene and conclude that the 1240 cm^{-1} band could be assigned to a C—H bending vibration.

The final resolution of the IR problem came from the group of Masamune[8] who used Fourier transform IR interferometry and found four bands below 2000 cm^{-1}. This was more than allowable for the D_{4h} structure. They further showed that the 1240 cm^{-1} band was indeed a C—H vibration by preparing cyclobutadiene-d_4 and observing a drop of nearly 200 cm^{-1} in the position of this band. It thus appeared that cyclobutadiene was not square after all, and that the spectra were consistent with a rectangular structure.

The cyclobutadiene saga serves as an illustration of the difficulty of assigning structures on the basis of IR spectroscopy of matrix-isolated intermediates.

8.2. MAGNETIC RESONANCE METHODS

8.2.1. NMR of Carbonium Ions in Superacid Media

Matrix isolation could be considered the most extreme technique for preventing reactive molecules from undergoing bimolecular reactions. It is essential if the reactive intermediate is prone to dimerization. For certain species, however, dimerization is not a problem and bimolecular reactions can be suppressed in another way. In the case of carbonium ions the partner in a bimolecular reaction would, by definition, be either a base or a nucleophile. By preparing carbonium ions in superacid media[9] such as FSO_3H or FSO_3H/SbF_5, one can ensure the absence of effective bases or nucleophiles. Under these circumstances many carbonium ions have long lifetimes, rendering them accessible to normal instrumental analysis techniques, especially NMR. One might question whether carbonium ions prepared in this way provide information that can be applied to solvolysis reactions where ion pairing and solvation are likely to be important[10], nevertheless, the technique has been useful in providing answers to a number of mechanistic problems,[11] not least of which is the nonclassical ion problem.

For more than two decades there was an argument among physical organic chemists about the existence of nonclassical carbonium ions. Few people doubted that such structures existed as transition states for the Wagner-Meerwein rearrangement but the problem was to determine whether they could exist as true intermediates, local minima on a potential energy surface. Much of the controversy centered on the 2-nonbornyl cation[12–14] but the definitive answer came from a study of cation 3 in a superacid medium.[15]

3c   3n

The delightful feature of cation 3 is that the classical ion 3c can show twofold symmetry (four resonances in the NMR) for a static structure or ninefold symmetry (one resonance) for rapidly equilibrating structures. The nonclassical trishomocyclopropenium structure 3n, on the other hand, has threefold symmetry (three resonances).

In fact the ion showed three resonances in both the ^{1}H and ^{13}C NMR, strongly supporting the nonclassical structure.[15] The procedure of assigning structure by counting peaks is much more reliable for NMR than IR spectroscopy for three reasons. First, all resonances in the NMR have intensities that can be estimated with some certainty from first principles. In the present case they are all expected to be of essentially equal intensity—there cannot be any small peaks hiding in the baseline noise as there can with IR spectroscopy. Second, the probability of accidental degeneracy of resonances in the ^{13}C NMR is very low because of the extremely narrow line width and large range of chemical shifts that occur in this type of spectroscopy. Third, the ability to look at the NMR spectra of different nuclei allows one to obtain independent sets of data that provide mutual corroboration.

The Coates' cation (3) is unusual in that simple counting of resonances allowed differentiation between classical and nonclassical structures. It is more usually the case that one predicts identical numbers of resonances for a nonclassical ion and a rapidly equilibrating set of its classical counterparts. Even in this situation one can sometimes still resolve the problem by applying a technique described by Saunders.[16] It is best illustrated by looking at the example of the 1,2-dimethylnorbornyl cation (4).

The ^{13}C NMR of 4 (R = H) shows a single line for C1 and C2 combined even at −150°C. Unfortunately this information does not allow one to distinguish between rapidly equilibrating classical structures (4c) and the symmetrical nonclassical structure (4n). By contrast, consider the situation that one would expect for the deuterated analog of 4 (R = D). The ^{13}C NMR should now show two resonances whose separation, δ, depends on the equilibrium constant, K, between the two classical structures and the intrinsic chemical shift difference, Δ, between C1 and C2 in a static classical ion. The relationship can be developed as follows. Let δ_1 and δ_2 be the chemical shifts of C1 and C2, repectively. (Note that the numerical labels are being applied to particular carbons, they are not being used in the IUPAC sense. Thus C1 is always the carbon bearing the CD_3.) If the mole fraction of ion A is χ then

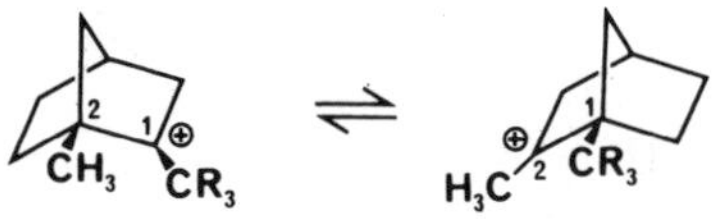

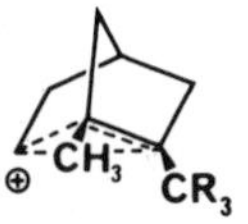

$$\delta_1 = \chi\delta_{C^0} + (1 - \chi)\delta_{C^.}$$

$$\delta_2 = \chi\delta_{C^.} + (1 - \chi)\delta_{C^0}$$

where δ_{C^0} and $\delta_{C^.}$ are the chemical shifts of bridgehead and cationic carbons in the static carbonium ion. Hence,

$$\delta = \chi\Delta - (1 - \chi)\Delta$$

Since

$$K = \frac{1 - \chi}{\chi}$$

then

$$\delta = \frac{1 - K}{1 + K} \cdot \Delta$$

If $R = H$ then $K = 1$ and so $\delta = 0$, that is, one sees a single line for C1 and C2 combined. However if $R = D$ then $K \neq 1$ and $\delta \neq 0$. A typical value of K might be 0.9, giving $\delta \simeq 0.05\Delta$. This might look very small until one recalls the large chemical shift dispersion in ^{13}C-NMR spectroscopy, especially for a carbonium ion where a value of $\Delta \simeq 200$ ppm would be quite reasonable. Thus a split of about 10 ppm in the C1 + C2 resonance could be expected upon deuteration if one were observing equilibrating classical ions.

How does this compare with the expectation for a single, nonclassical structure? Saunders reasoned that the effect of deuteration on the nonclassical ion would be similar to that on an allyl cation. He chose the 3-cyclohexenyl cation as a convenient example and found that the addition of one deuterium caused a split of < 0.5 ppm.[16] Thus there ought to be something between one and two orders of magnitude difference in the splitting of the ^{13}C-NMR signal for the classical and nonclassical ions.

Experimentally Saunders found $\delta = 20.5$ ppm at $-170°$C for the 1,2-dimethyl-norbornyl cation, clearly indicating equilibrating classical structures. By contrast the parent norbornyl cation exhibited $\delta < 2$ ppm (line broadening due to the 6,2 hydride shift prevented complete resolution), which Saunders took to be support for a nonclassical structure.

The Saunders' technique need not be restricted to carbonium ions although the smaller value of Δ expected for uncharged molecules makes interpretation of the results somewhat more difficult. Maier and coworkers[17] recently used it to demonstrate equilibration between nonsquare isomers of tri-*tert*-butylcyclobutadiene.

8.2.2. Chemically Induced Dynamic Nuclear Polarization

In the strictest sense the phenomenon of chemically induced dynamic nuclear polarization (CIDNP) does not correspond to a *direct* detection of a reactive intermediate since it is expressed as unusual features in the *product* NMR spectra. It is included in this chapter because it is an instrumental technique that involves the monitoring of an on-going reaction and it is, in that sense, related to the other techniques discussed here.

CIDNP is a phenomenon that occurs during reactions involving the formation of radical pairs in a solvent cage. It consists of a sorting of nuclear spins (usually of hydrogen atoms) between products derived from cage escape and those from geminate recombination or disproportionation within the cage. It is detected as enhanced absorption or emission (inverted peaks) in the NMR spectra of the products. Two subclassifications of CIDNP are recognized. They are commonly called the "net" effect and the "multiplet" effect. We will consider only the first in detail, since it is the more common. The reader interested in a more complete description should read the papers by Kaptein.[18]

Consider the reaction of generalized molecule **5** to give a radical pair in a solvent cage. Let us suppose, for the moment, that the radical pair is formed with spin conservation, in other words, that it is in a singlet spin state. Let us further suppose that the radical bearing the hydrogen atom has the larger g factor. (The g factor is a measure of the local magnetic field experienced by the unpaired electron. It can be considered roughly analogous to the chemical shift parameter in NMR spectroscopy.) The initial singlet state of the radical pair can be represented by the vector diagram shown below in which the arrows represent the

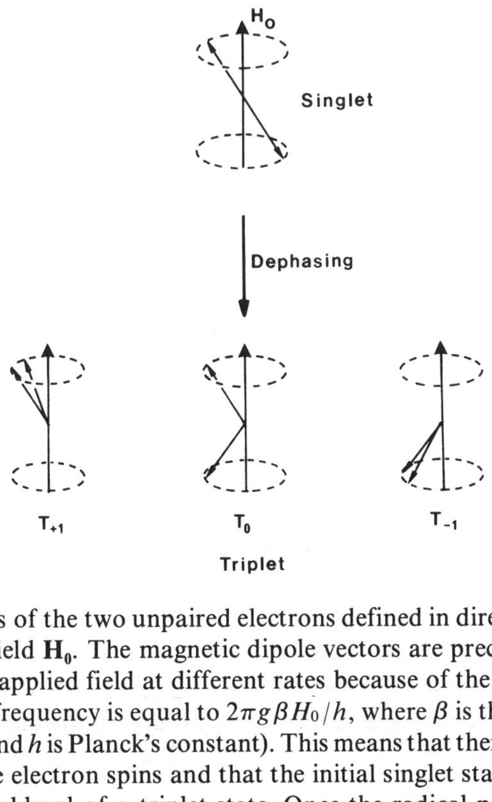

H_0

Singlet

Dephasing

T_{+1} T_0 T_{-1}

Triplet

magnetic dipoles of the two unpaired electrons defined in direction with respect to an external field H_0. The magnetic dipole vectors are precessing around the direction of the applied field at different rates because of the different g factors (the precession frequency is equal to $2\pi g \beta H_0/h$, where β is the Bohr magneton for an electron and h is Planck's constant). This means that there will be a gradual dephasing of the electron spins and that the initial singlet state will, given time, become the T_0 sublevel of a triplet state. Once the radical pair has achieved a triplet state it can no longer undergo geminate recombination or disproportionation and so all such radical pairs suffer cage escape.

The hydrogen nuclear spin enters the picture through its influence on the dephasing rate. If the hydrogen nucleus has an α spin it will cause a small preference for the electron nearer the hydrogen to have an α spin. Since the electrons making up the C—H σ bond have antiparallel spins, the electron of the C—H bond nearer the carbon will have a slightly higher probability of possessing a β spin. This means that the unpaired electron on carbon will experience a somewhat reduced net magnetic field, resulting in a lower precession frequency. Remembering that we assigned the larger g factor to the radical bearing the hydrogen atom, we can see that the final result will be a decrease in the rate of singlet→triplet dephasing and an increase in the probability of geminate recombination or disproportionation within the solvent cage.

In summary, then, one is led to conclude, for this example, that the products of reactions occurring within the solvent cage will have hydrogens with a slight excess of α spins. By difference the products derived from cage escape will have hydrogens with an excess of β spins. Since the α spin state is of lower energy the experimental result is that the ^{1}H-NMR spectrum of the products from reaction within the solvent cage exhibit enhanced absorption for the hydrogen in question whereas the corresponding NMR resonance for the products of cage escape appears in emission (inverted).

Several features were assumed for this example. A change in any one of them would have reversed our final conclusion about which products exhibit enhanced absorption and which exhibit emmission. Had the original radical pair been formed in a triplet state, the decrease in dephasing rate caused by the hydrogens with α spin would have led to a *decrease* in the proportion of geminate recombination. If the hydrogen had been attached to the radical with the *smaller g* factor in a singlet radical pair, the reduction in precession frequency caused by the hydrogens with α spin would have led to an *increase* in the dephasing rate and, again, a *decrease* in the proportion of geminate recombination. Finally, if the hydrogen under observation had been one bond further removed from the radical center its coupling constant would have changed from negative to positive and the influence of α and β nuclear spins would have been reversed. These factors are summarized in a rule proposed by Kaptein.[18]

$$\Gamma_n = \mu\epsilon \, \Delta g \, a_i$$

μ is positive for a radical pair in an initial triplet state.

ϵ is positive if the product under observation arose from reactions within the solvent cage.

Δg is the sign of $(g_i\text{-}g)$ where g_i is the g factor of the radical bearing the hydrogen under observation and g is the g factor of the other radical.

a_i is the sign of the coupling constant from the hydrogen under observation to the unpaired electron.

If Γ_n is positive one will observe enhanced absorption (A) whereas if Γ_n is negative one will see an emission signal (E).

Note that net effect CIDNP can be observed only if the two radicals have different g factors. The multiplet effect can lead to CIDNP from a pair of identical radicals under special circumstances[18] but this is not of great importance for the present purposes.

The discussion in this section has focused on CIDNP in ^{1}H NMR but it should be clear from the analysis that any NMR active nucleus can exhibit the phenomenon in principle.

One might think that CIDNP would be a useful criterion for the existence of radical pairs as reaction intermediates but this really is not the case. The problem is that neither the observation of CIDNP nor the failure to observe it tells one anything conclusive about the major reaction pathway. The disturbance in the population of spins caused by CIDNP is so large compared to the normal population difference between high and low energy states in an NMR experiment that the signals tend to swamp out the normal resonances of the products. As a consequence it is possible to obtain strong CIDNP signals from radical pairs that constitute only a few percent of the total reaction. On the other hand there can be circumstances where the major reaction pathway does involve radical pairs but no CIDNP signal is observed because all reactions occur within the solvent cage.

The real strength of CIDNP is its ability to provide information about the nature of the radical pairs in a reaction where independent evidence has established that a radical pathway is dominant. An interesting example is provided by the work of Bartlett and Shimizu on the thermolysis of benzoyl peroxide.[19]

Direct photolysis of benzoyl peroxide in carbon tetrachloride results in the formation of chlorobenzene in which the *ortho* hydrogens appear in emission. This is the expected result for a reaction involving the intermediacy of a singlet radical pair:

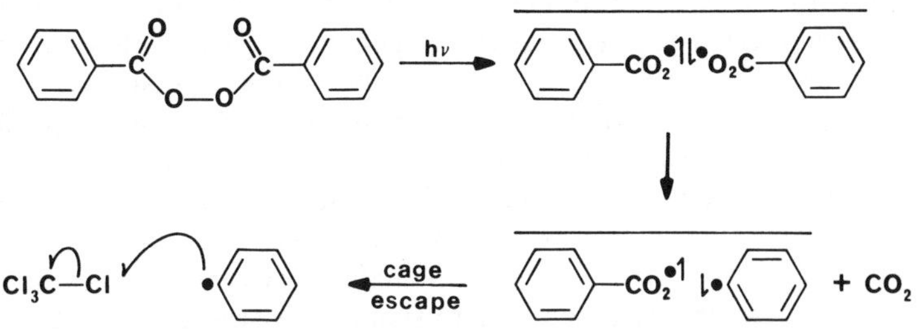

The first formed radical pair has $\Delta g = 0$ and so could exhibit no net effect CIDNP. After decarboxylation of one radical, however, g_r-$g > 0$ (the oxygen-centered carboxyl radical will have the larger g factor) and so CIDNP becomes possible. Assigning $\mu = -$ for the singlet radical pair, $\epsilon = -$ for observation of a product formed by cage escape, $\Delta g = -$ as discussed above, and $a_i = +$ for the observation of hydrogens β to the radical center, one finds $\Gamma_n = -$, leading to the expectation of an emission signal, in accord with the observation.

By contrast, the thermolysis of dibenzyl peroxide in carbon tetrachloride in the presence of tetramethyldioxetane results in the formation of chlorobenzene in which the *ortho* hydrogens exhibit enhanced absorbtion. Given that ϵ, Δg, and a_i must have the same signs as in the previous experiment, the only explanation seems to be that μ has changed sign, in other words that the radical pair is now formed in an initial triplet state. At first sight this might appear implausible but, in fact, it is known that tetramethyldioxetane does undergo thermal decomposition to give one equivalent of acetone in its $n\pi^*$ triplet state.[20] It seems likely, therefore, that the mechanism of the present reaction involves triplet sensitization of the benzoyl peroxide decomposition.

8.2.3. Spin-Saturation Transfer

The technique of spin-saturation transfer was first applied to continuous wave ^{1}H NMR by Forsén and Hoffman.[21] A later modification by these authors[22] and by Mann[23] allowed its application to Fourier transform ^{13}C NMR, thereby greatly improving its utility.

The technique involves the irradiation of specific resonances in the NMR spectrum of a system that is undergoing reaction but is still in the slow exchange limit. The irradiation causes saturation of the spins of the selected nuclei. This saturation is transferred to different sites in the molecule by the reaction under study and is recognized as a decrease in magnetization (signal intensity) of those sites. Two limits are recognized: if τ (the lifetime of the irradiated nucleus at its original site) is $\ll T_1$ (the spin-lattice relaxation time of the nucleus at the new, nonirradiated site) then spin-saturation transfer is essentially complete and the signals for the new sites will disappear. If, on the other hand, $\tau \gg T_1$ then no effective spin-saturation is transferred and the signals for the new sites will be unaffected. Of special interest is the intermediate case where $\tau \simeq T_1$. Under these circumstances there will be a decrease in signal intensity for the new sites but not a complete disappearance of the resonance. By measuring the decrease in signal intensity and independently determining T_1, it is possible to calculate τ, or τ^{-1}, which is the rate constant for the site exchange process.

The technique provides some of the same information as an isotopic labeling experiment but it has some advantages and some disadvantages. The advantages are, first, that it is not necessary to synthesize an isotopically labeled reactant and, second, that the intermediate case of $\tau \simeq T_1$ allows one to determine rate constants. The big disadvantage is that the rate constant must be of such a magnitude (typically 1–10 $\sec^{-1}$ for ^{13}C NMR) that the technique can really be applied with ease only to rearrangements where both reactant and product have substantial concentrations at equilibrium. Irreversible reactions would be essentially complete before meaningful measurements could be taken. The limitations on the magnitude of the rate constant also represent a constraint on the types of reaction that can be studied by this method. While it is possible to adjust τ by changing the temperature, the temperature required for reactions with large activation free energies might be inaccessibly high.

Spin-saturation transfer has been used to determine the activation parameters for rearrangement of carbonium ions such as various barbaralyl cations[24] and to determine the mechanisms of degenerate rearrangements such as the metal migration in $[\eta^6-C_8H_8]Cr(CO)_3$. In the latter example Mann[25] was able to show that the rearrangement occurred by a combination of 1,2 and 1,3 shifts and did not involve the 20 electron $[\eta^8-C_8H_8]Cr(CO)_3$ structure that had been previously postulated. Had such an intermediate been involved, the spin saturation caused by irradiating any one pair of carbons would have been transferred randomly to all of the other carbons. This did not occur.

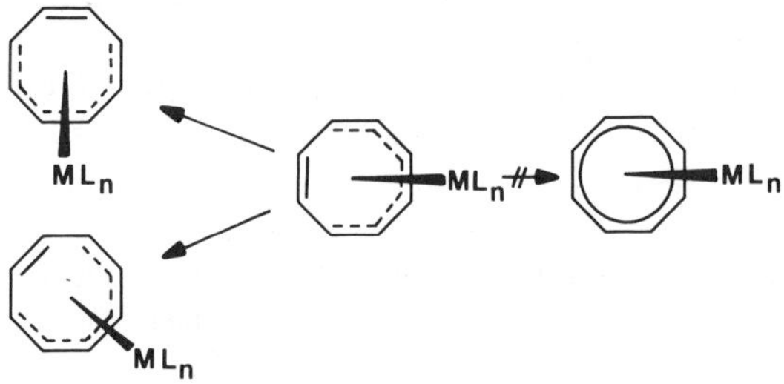

8.2.4   Electron Spin Resonance

Electron spin resonance (ESR) spectroscopy is a technique that relies on the presence of unpaired electrons in a molecule. In the case where the molecule is a free radical (one unpaired electron) the phenomenon can be considered to be analogous to ^1H or ^{13}C NMR except that it is a change in spin state of the odd electron rather than a change in nuclear spin that is responsible for the absorption of electromagnetic radiation. In just the same way that the chemical shift of some particular nucleus is a reflection of the local magnetic field experienced by that nucleus, so the "g factor" of a radical reflects the local magnetic field experienced by the odd electron. The coupling of nuclear spins in NMR spectroscopy also finds a parallel in the "hyperfine coupling" between the electron spin and the spins of nearby nuclei (especially ^1H, ^{13}C, and ^{14}N). A number of texts have been written on the topic of ESR spectroscopy. One that covers most of the theoretical aspects and a number of experimental applications is the book by Ayscough.[26]

There is some hazard in using the presence of an ESR signal as an indication of the involvement of radicals in a given reaction for essentially the same reason that the presence of CIDNP can be misleading. The technique is very sensitive and gives no easy method of calibration that would allow one to estimate the fraction of the reaction that is proceeding by a radical pathway. ESR is much more useful for determining something about the properties and reactions of radicals. Unimolecular rearrangements of radicals can be followed by ESR and can provide information that would be difficult to obtain any other way.[26]

Much recent interest has focused on the ESR spectroscopy of biradicals and carbenes in triplet states. The observation of these spectra requires that the sample be immobilized in a single crystal or in a glassy matrix (often 2-methyltetrahydrofuran). The spectra can reveal a number of useful pieces of information about the species under observation: the "zero field splitting" parameters D and E provide information about the degree of interaction between the unpaired electrons and the symmetry of the molecule (the E value is identically 0 for triplet state molecules with threefold or higher symmetry).[27] Temperature dependence studies can allow one to determine whether the molecule under observation has a triplet groundstate or a singlet groundstate with a thermally populated triplet state. It is sometimes possible to determine the singlet-triplet energy difference[28] provided that it is not $\gg$ RT. Recent examples of mechanistic interesting studies on triplet state molecules come from the work of Dowd, Platz, and Berson.

Dowd was the first to report the ESR spectrum of the biradical trimethylene-methane (6) in an oriented single crystal. From the zero value of the E parameter it was possible to deduce that the molecule had threefold symmetry[29] and from the fact that the signals obeyed the Curie law (signal intensity varies linearly with $1/T$), it could be deduced that the triplet was the groundstate (and that no singlet states were thermally accessible).[30] More recently Dowd's group has investigated the ring closure of trimethylenemethane to methylene cyclopropane.[31] They find

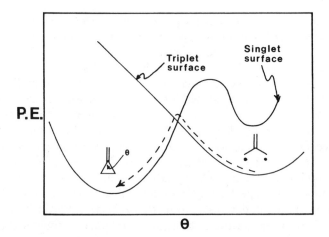

6

an activation barrier of 7 kcal/mol. If one assumes that the observed triplet state of 6 must intersystem cross to a singlet before ring closure can occur then the activation energy can be taken as an upper bound on the singlet–triplet energy difference (Δ_{ST}). This approach suggests a much smaller value for Δ_{ST} than that obtained from quantum mechanical calculations ($\approx$20 kcal/mol). One possible explanation is that the ring closure is not spin conservative. In other words that the singlet and triplet surfaces intersect at an energy below that of the first singlet state of 6. The large negative activation entropy observed for the reaction ($\Delta S^{\ddagger} = -17$ cal mol^{-1} K^{-1}) would be consistent with such an interpretation.

Unfortunately this does not completely resolve the disagreement between theory and experiment because an extensive calculation by Davidson, Borden, and coworkers[31] finds that the intersection between singlet and triplet surfaces occurs at higher energy than the first singlet state of **6**.

Platz[32] has used ESR spectroscopy to show that diazo-compounds such as **7** undergo photolysis to give biradicals, presumably by intramolecular H abstraction from the carbene. Related intermolecular H abstractions were studied[33] and shown to involve considerable contributions from quantum mechanical tunneling.

7

Berson[34] has used ESR spectroscopy to investigate the chemistry of *m*-napthoquinomethane (**8**).

8

It was found that irradiation of 6-methylenebenzobicyclo [3.1.0] hexen-2-one resulted in the appearance of a triplet signal in the ESR. This signal showed a zero field splitting parameter, D, that was close to the one estimated for **8** from semiempirical molecular orbital calculations. The signal intensity did not show strict Curie law adherence, but it could not be determined whether the curvature was due to the presence of a nearby singlet state or whether it was an artifact due to saturation effects. If the former was assumed, it could be shown that the hypothetical singlet state could be no more than 0.01 kcal/mol below the observed triplet. This result is of interest; while a triplet groundstate is expected for the hydrocarbon *m*-quinodimethane one might have guessed that the oxygen analog **8**, lacking anything but accidental orbital degeneracy, would have had a strong preference for the singlet configuration. The contrary experimental evidence supports a theoretical classification of biradicals suggested by Ovchinnikov.[35]

8.3. LASER TECHNIQUES

Laser radiation has a number of properties that make it especially well suited to chemical studies. It is intense, highly monochromatic, coherent, frequently tunable (with the spectrum from IR to vacuum UV being essentially completely covered), and can be produced in pulses of extremely small duration. These properties have led to a rapid escalation of the use of lasers for studying a wide variety of fundamental chemical transformations. In this section we will discuss some applications that have the most direct relevance to organic reaction mechanisms.

8.3.1. Picosecond Spectroscopy

Advances in laser technology have made the generation of picosecond (1 psec $= 10^{-12}$ sec) pulses of light an almost routine procedure. These pulses allow the detection of extremely short-lived reactive intermediates. The only restriction is that the intermediates must be generated photochemically. Studies of multi-photon dissociation using IR lasers (which would better mimic thermal reactions) have been done but the time scales are necessarily much longer. We will consider one recent study, conducted in solution, as an illustration of the technique.

In Chapter 5 (Section 5.4) we considered a redox reaction involving formal transfer of a hydride from a NADH model to a ketone. Isotope effect studies were consistent with a single step reaction but could not rule out the existence of an intermediate such as a radical ion pair created by single electron transfer. A similar reaction has been studied by picosecond laser spectroscopy, with one important difference: the ketone was converted to its first excited triplet state (in a sense photooxidized since the electronic excitation opens a "hole" in a low-energy molecular orbital) before the reaction occurred. This study does not have any bearing on the question of whether there is an intermediate in the thermal reaction but can provide information about the steps that would follow an initial electron transfer. The reducing agent in the study by Peters and coworkers[36] was *N*-methylacridan (NMA), **9**, while the oxidizing agent was triplet benzophenone, **10**.

Irradiation of the sample at 355 nm resulted in formation of T_1 benzophenone within 10 psec. This state was quenched within 25 psec using $1 M$ NMA in benzene. Concomitant with the quenching was the appearance of a new state having absorption maxima at 640 and 720 nm. The 720-nm absorption was known from previous studies to correspond to the benzophenone radical anion, hence the

640-nm absorption could be ascribed to the NMA radical cation. The 720-nm absorption was found to shift to 690 nm with a half life of 100 ± 50 psec. Again this behavior had been seen in earlier studies and was ascribed to conversion of an initial solvent-separated ion pair to a contact ion pair. The next step, occurring with a half-life of 500 ± 100 psec, consisted of proton transfer to give the known benzophenone ketyl radical (λ_{max} 545 nm) and the N-methylacridinium radical (λ_{max} 525 nm). The final step observed in this study was a second electron transfer to give hydroxydiphenylmethyl carbanion (λ_{max} 450 nm) and N-methylacridinium ion.

At first sight it might appear that a thermal reaction involving single electron transfer would intersect the sequence outlined above at the formation of the solvent-separated radical ion pair. In fact this is not strictly correct. The radical ion pair formed in this study presumably has a net triplet state. It is prohibited from undergoing hydrogen atom transfer (which would give hydroxydiphenyl-methyl anion and acridinium ion directly) because this would require the energetically unfavorable formation of triplet N-methylacridinium. A singlet-state radical ion pair formed in a thermal reaction would have no such prohibition.

Peters and coworkers addressed this problem by studying the NMA/fluorenone reaction. Fluorenone does not intersystem cross on the time scale of the reaction and so all of the chemistry takes place in the singlet manifold, as it would for a thermal reaction. Remarkably, even with fluorenone as the hydrogen acceptor, the spectroscopic study still showed evidence for proton rather than hydrogen atom transfer within the radical ion pair.

8.3.2. Laser Induced Fluorescence

One of the difficulties of flash photolysis experiments such as that described above is that one has to identify transients on the basis of a single datum, the absorption maximum in the electronic spectrum. In order to be secure in one's assignments it is thus necessary to perform a number of control experiments in which the putative reactive intermediates are generated from a variety of different sources. A technique that promises to overcome this problem, at least for gas phase studies, is laser induced fluorescence (LIF).[37] In this technique the reactive intermediate, generated by a pulse from a pump laser, is excited to a higher-lying electronic state by a second tunable laser and is then detected by measurement of its fluorescence spectrum. The important feature of the technique is that the excitation spectrum (fluorescence intensity vs. probe laser wavelength) has vibrational and rotational fine structure that provides an unambiguous "fingerprint" for the transient molecule. In addition the intensities of the various transitions provide information about the population of vibrational and rotational states of the nascent reactive intermediate. In reactions that are state-selective this can provide invaluable details of the reaction mechanism.

Zare and coworkers[38] used LIF to show that singlet ($^1A_1$) methylene was produced by photolysis of ketene at 337.1 nm. The excitation spectrum showed that the methylene was in its ground vibrational state and in rotational states up to $J = 4$. From the known heats of formation of ketene and carbon monoxide and from the known energy content of the photon from the pump laser it was possible to deduce the heat of formation of $^1A_1$ methylene in its lowest vibrational and rotational states. This value (101.7 ± 0.5 kcal/mol) could be compared with the previously determined heat of formation for groundstate methylene ($^3B_1$; 93.6 ± 0.6 kcal/mol) to obtain the adiabatic singlet–triplet energy separation of 8.1 ± 0.8 kcal/mol, a number that had been the source of much controversy between experimentalists and theoreticians for many years before Zare's work.

REFERENCES

1. C. Y. Lin and A. Krantz, *J. Chem. Soc. Chem. Commun.*, 1111 (1972).

2. O. L. Chapman, C. L. McIntosh, and J. Pacansky, *J. Am. Chem. Soc.*, **95**, 614 (1973).

3. G. Maier, H. -G. Hartan, and T. Sayrac, *Angew. Chem. Int. Ed. Engl.*, **15**, 226 (1976).

4. R. G. S. Pong, B. -S. Huang, J. Laureni, and A. Krantz, *J. Am. Chem. Soc.*, **99**, 4153 (1977).

5. A. Krantz, C. Y. Lin, and M. D. Newton, *J. Am. Chem. Soc.*, **95**, 2744 (1973).

6. H. Kollmar and V. Staemmler, *J. Am. Chem. Soc.*, **100**, 4304 (1978).

7. L. J. Schaad, B. A. Hess, Jr., and C. S. Ewig, *J. Am. Chem. Soc.*, **101**, 2281 (1979).

8. S. Masamune, F. A. Souto-Bachiller, T. Machiguchi, and J. E. Bertie, *J. Am. Chem. Soc.*, **100**, 4889 (1978).

9. R. J. Gillespie, *Acc. Chem. Res.*, **1**, 202 (1968).

10. See the discussion of solvolysis in Chapter 4.

11. G. A. Olah and J. A. Olah, in *Carbonium Ions*, Vol. II, G. A. Olah and P. von R. Schleyer (Eds.), Wiley, New York, 1970, p. 715.

12. S. Winstein and D. S. Trifan, *J. Am. Chem. Soc.*, **71**, 2953 (1949).

13. H. C. Brown, I. Rothberg, P. von R. Schleyer, M. M. Donaldson, and J. J. Harper, *Proc. Natl. Acad. Sci.*, **56**, 1653 (1967).

14. E. M. Arnett, N. Pienta, and C. Petro, *J. Am. Chem. Soc.*, **102**, 398 (1980).

15. R. M. Coates and E. R. Fretz, *J. Am. Chem. Soc.*, **99**, 297 (1977).

16. M. Saunders, L. Telkowski, and M. R. Kates, *J. Am. Chem. Soc.*, **99**, 8070 (1977); M. Saunders and M. R. Kates, *Ibid.*, **99**, 8071 (1977).

17. G. Maier, H.-O. Kalinowski, and K. Euler, *Angew. Chem. Int. Ed. Engl.*, **21**, 693 (1982).

18. R. Kaptein, *J. Chem. Soc. D*, 732 (1971); R. Kaptein, *Adv. Free Radical Chem.*, **5**, 319 (1975).

19. P. D. Bartlett and N. Shimizu, *J. Am. Chem. Soc.*, **97**, 6253 (1975).

20. T. Wilson, M. E. Landis, A. L. Baumstark, and P. D. Bartlett, *J. Am. Chem. Soc.*, **95**, 4765 (1973).

21. S. Forsén and R. A. Hoffman, *Acta Chem. Scand.*, **17**, 1787 (1963); *Prog. NMR Spec.*, **1**, 173 (1966).

22. S. Forsén and R. A. Hoffman, *J. Chem. Phys.*, **40**, 1189 (1964).

23. B. E. Mann, *J. Mag. Res.*, **21**, 17 (1976); **25**, 91 (1977).

24. C. Engdahl and P. Ahlberg, *J. Am. Chem. Soc.*, **101**, 3940 (1979).

25. B. E. Mann, *J. Chem. Soc. Chem. Commun.*, 626 (1977).

26. P. B. Ayscough, *Electron Spin Resonance in Chemistry*, Methuen, London, 1967.

27. E. Wasserman, L. C. Snyder, and W. A. Yager, *J. Chem. Phys.*, **41**, 1763 (1964).

28. See, for example, reference 26, p. 417.

29. P. Dowd and K. Sachdev, *J. Am. Chem. Soc.*, **89**, 715 (1967).

30. P. Dowd, *Acc. Chem. Res.*, **5**, 242 (1972).

31. P. Dowd and M. Chow, *J. Am. Chem. Soc.*, **99**, 6438 (1977); D. Feller, K. Tanaka, E. R. Davidson, and W. T. Borden, *Ibid.*, **104**, 967 (1982).

32. M. S. Platz, *J. Am. Chem. Soc.*, **102**, 1192 (1980).

33. M. S. Platz, V. P. Senthilnathan, B. B. Wright, and C. W. McCurdy, Jr., *J. A.n. Chem. Soc.*, **104**, 6494 (1982).

34. D. E. Seeger, E. F. Hilinski, and J. A. Berson, *J. Am. Chem. Soc.*, **103**, 720 (1981).

35. A. A. Ovchinnikov, *Theor. Chim. Acta*, **47**, 297 (1978).

36. K. S. Peters, E. Pang, and J. Rudzki, *J. Am. Chem. Soc.*, **104**, 5535 (1982).

37. J. L. Kinsey, *Ann. Rev. Phys. Chem.*, **28**, 349 (1977).

38. R. K. Lengel and R. N. Zare, *J. Am. Chem. Soc.*, **100**, 7495 (1978).

CHAPTER 9

DETAILED CASE HISTORIES

The task of determining a plausible reaction mechanism in a real research situation has three main phases. First, one must be able to think of at least one mechanism to be tested. Second, one must design an experiment or set of experiments to test the mechanistic hypothesis. Third, one must execute these experiments. As we saw in Chapter 1, it may be necessary to repeat this sequence several times before a "satisfactory" mechanism can be proposed.

Stages 1 and 2 typically take somewhere between a few minutes and a few days. Stage 3 typically takes a few months to a few years! In Chapters 1–8 we have focused on stage 1 and, especially, stage 2, but it is the realization that stage 3 is to come that must inevitably influence one's selection of technique and, where appropriate, substrate. For example, it is not sufficient to recognize that some experiment involving optically active, stereospecifically labeled reactants would serve to distinguish between two mechanisms *in principle*. One must also be concerned about how such compounds would be synthesized and whether it would be possible to analyze the products to obtain the necessary information. There is an all-important trade-off between the abstract desire for elegance and the pragmatic requirement of finishing the experiment in a finite time. In the modern research environment this conflict of interests is heightened by the fact that it is frequently different individuals who are responsible for the design and the execution of experiments!

In this final chapter we will look at some case histories that serve to illustrate many of the techniques that were discussed in the previous pages and, in some cases, highlight the problem of how one obtains rigorous and significant data with practicable experiments.

9.1. THE VON RICHTER REACTION

The remarkable von Richter reaction is a nice example of the cycle through hypothesis and experiment that was discussed in Chapter 1. It also serves as a sobering reminder that one never knows for sure when this cycle has come to an end!

In 1871 von Richter reported that nitrobenzene reacted with potassium cyanide in aqueous ethanol at high temperature to give benzoic acid and potassium nitrite. He further discovered that p-substituted nitrobenzenes gave m-substituted benzoic acids in 5–50% yield along with substantial quantities of "acidic tars."

The first suggestion of a mechanism for the reaction came from Bunnett and coworkers[1] 83 years later! They noted that the apparent migration of a substituent from the *para* to *meta* site was inconsistent with a direct nucleophilic sub-

stitution. The same observation made a benzyne mechanism seem implausible since it might have been expected to give both *meta* and *para* products. The possibility of preferential *meta* attack on a benzyne was ruled out by observing that m-substituted nitrobenzenes gave no *meta* products. A plausible mechanism thus seemed to be a cine substitution in which cyanide ion acts first as a nucleophile then as a base to expel nitrite ion. This mechanism predicts the

incorporation of one deuterium *ortho* to the carboxylic acid residue if the reaction is run in $D_2O/EtOD$. When the experiment was carried out a total of 0.72 D per molecule was incorporated but at unknown sites. In order to determine the location of the isotope exchange Bunnett and coworkers prepared the dideuterated substrate 1 and found that the product of the von Richter reaction contained 0.8 D per molecule, in reasonable accord with expectation.

The first doubts came 2 years later when Bunnett and coworkers[2] found that 2-nitronaphthalene gave naphthalene-1-carboxylic acid with potassium cyanide in refluxing ethanol/water but naphthalene-1-carbonitrile, a supposed intermediate in the reaction, would not hydrolyze under the conditions. This result proved conclusively that their mechanism could not apply to the naphthalene version of the von Richter reaction and, by virtue of Occam's razor, called it into question for the benzene cases as well. Returning to the *p*-chloronitrobenzene substrate, they found that *m*-chlorobenzoic acid was produced under the von Richter conditions at 150° C but that *m*-chlorobenzonitrile would only hydrolyze as far as the amide in the same medium. This showed that neither the nitrile nor the amide could be an intermediate in the von Richter reaction and that their mechanism was, indeed, incorrect.

Their task now was to propose a mechanism that would incorporate the previous data concerning deuterium incorporation and substituent migration without involving a benzonitrile or benzamide derivative. The one they came up with is shown in Figure 9.1.

This mechanism made the prediction that one of the oxygens in the final carboxyl group would come from the original nitro group and one from the aqueous solvent, provided that the intermediate mixed anhydride hydrolyzed by O—N rather than O—C bond cleavage. The prediction was verified by Samuel[3] in 1960. He ran the von Richter reaction of *p*-bromonitrobenzene in water containing 1.53% ^{18}O. The product, *m*-bromobenzoic acid, was found by mass spectrometry to contain 0.84% ^{18}O. Since the natural abundance of ^{18}O is 0.20%, the expected ^{18}O content for a product incorporating just one solvent oxygen

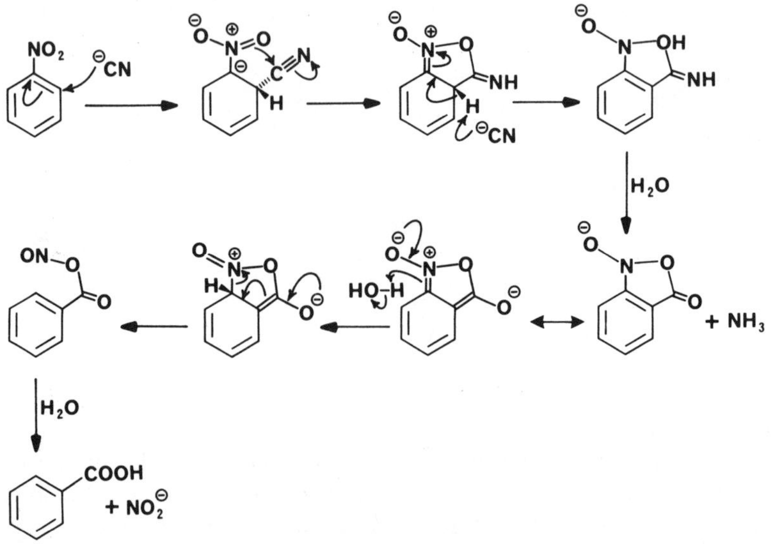

FIGURE 9.1.   A mechanism for the von Richter reaction that would obviate the need to involve the aromatic nitrile or amide.

would be 0.865%. This aspect of the Bunnett mechanism was thus verified. But the complete mechanism was still not correct!

In the same year that Samuel conducted his experiments Rosenblum[4] reread the original von Richter paper and discovered that the reported formation of nitrite ion was "deduced" from the stoichiometry but was not actually demonstrated. When Rosenblum conducted a quantitative analysis for NO_2^- in the von Richter reaction of p-chloronitrobenzene he found <3% based on the yield of m-chlorobenzoic acid. What he did find was molecular nitrogen formed in essentially equivalent amount to the m-chlorobenzoic acid. This discovery need not have ruled out the Bunnett mechanism because ammonia and nitrous acid react to give ammonium nitrite, which does then give molecular nitrogen upon heating. However Rosenblum added $^{15}NH_3$ to an on-going reaction and found that the N_2 product was all $^{14}N\equiv^{14}N$. This did rule out the Bunnett mechanism.

In order to ascertain the source of the nitrogen Rosenblum conducted another labeling experiment. p-Chloronitrobenzene labeled with $9.0 \pm 0.2\%$ ^{15}N was allowed to react with potassium cyanide containing only natural abundance (0.37%) ^{15}N. The nitrogen product was collected and analyzed by mass spectrometry, which revealed 9.27% mass 29. This showed that one atom of the N_2 came from the nitro group and one from the cyanide. Now it was time to propose a new mechanism that incorporated all of the facts including the nature of the products, the migration of the substituent and the ^2H, ^{15}N, and ^{18}O labeling studies. The mechanism shown in Figure 9.2 was suggested by Woodward.

Two subsequent experiments have supported Woodward's mechanism. In 1962 Ullman and Bartkus[5] treated the acyl hydrazide 2 with lead tetraacetate and

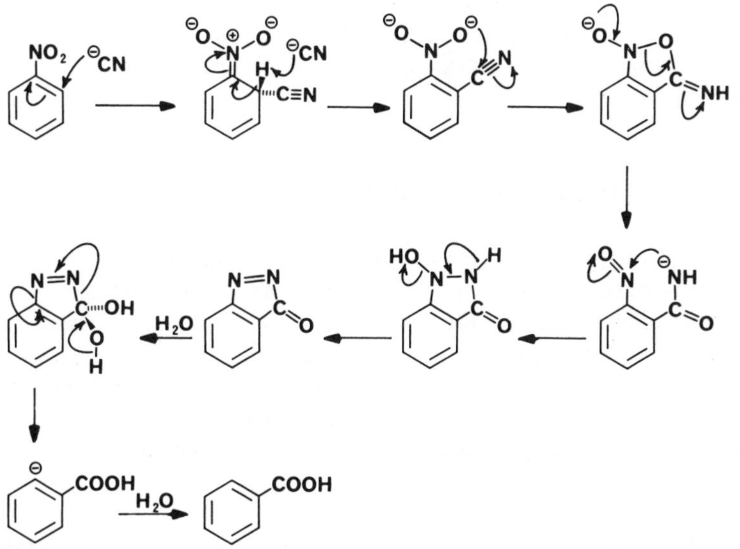

FIGURE 9.2. The presently accepted mechanism for the von Richter reaction.

obtained an intermediate with a deep red color. This intermediate would undergo cycloaddition to cyclopentadiene to give **4**, strongly suggesting that it was the diazene **3** proposed as an intermediate in Woodward's mechanism. Significantly, treatment of the red intermediate with KCN in aqueous ethanol resulted in formation of benzoic acid and nitrogen.

A year later Ibne-Rasa and Koubek[6] showed that a second intermediate in the Woodward mechanism, o-nitrosobenzamide, would go on to the products under conditions of the von Richter reaction. In addition they noticed the appearance of a transient deep red color in the reaction, in accord with the proposed involvement of **3**.

Woodward's mechanism is still the accepted one for the von Richter reaction, although, given the history of the problem, one hesitates to call it the "correct" mechanism!

9.2. THE [1,5] HYDROGEN MIGRATION

The study of the [1,5] hydrogen migration by Roth and coworkers[7] provides a nice example of the combined use of chirality and isotopic labeling. In the design of the substrate one also sees the balance between elegance and practicality that was discussed in the introduction to this chapter.

The problem that Roth wished to address was the stereochemistry of the [1,5] H migration in acyclic dienes. The Woodward–Hoffmann rules would mandate a suprafacial path but there was, at the time, no experimental evidence on this point.

A. The Strategy.   The difference between the suprafacial and antarafacial pathways is clearly the stereochemistry at the carbon bearing R_1 and R_2. It is immediately clear, then, that this is a chirality problem. It further follows that if

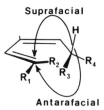

Suprafacial

Antarafacial

the center is to be chiral in the product R_1 *and* R_2 *must be different and neither can be H.* A further restriction is that one does not want R_2 to be an alkyl group because this would allow for the possibility of an unwanted [1,5] H migration:

A suitable solution would be $R_1 = CH_3$, $R_2 = D$. A more "elegant" solution would be $R_1 = T$, $R_2 = D$ but this would present severe synthetic and analytical problems. So, then, is **5** a suitable substrate for the study? A few moments

5

thought show that the answer is "no." The free rotation about the C—CH$_3$ bond allows both suprafacial and antarafacial pathways to give the same product. In order to determine which rotamer gave which product, one must make the

Suprafacial   **Antarafacial**

carbon bearing R_3 and R_4 chiral in the starting material. This means that R_3 *and* R_4 *must be different and neither can be H.* This time hydrogen isotopes are undesirable as choices for R_3 and R_4 because their migration in competition with the H would lead to a very complex product mixture that would be extremely difficult to analyze. A suitable solution is R_3, $R_4 = CH_3$, C_2H_5 (compound **6**). An added benefit of having two alkyl groups on this carbon is that the equilibrium constant will now favor the product of the [1,5] H migration:

Now that we have decided on the chemical constitution of the reactant we must decide what its chiroptical properties are to be. Thus the migration origin is chiral but does the molecule have to be optically active? The answer is "yes" because the products of suprafacial and antarafacial migration from a given rotamer are related as enantiomers. Had the reactant been racemic they would have been formed in equal proportions regardless of the mechanism.

So the substrate must be optically active but is it necessary to relate the configurations of reactant and products? Regrettably (because it is a lot of work) the answer is "yes." Neither mechanism would give racemic products from an optically active reactant.

In summary, then, one must devise a synthesis of 6 in optically active form and one must relate its absolute configuration to the products 7 and 8.

B. The Synthesis. The choice of synthetic strategy for Roth and coworkers was influenced by the fact that 9 and 10 in optically active form were already known. In the case of 9 the absolute configuration was also known. The researchers actually devised routes to 6 from both 9 and 10. We will concentrate on the former.

The retrosynthetic analysis from 6 is directed in part by the need to use 9 as a starting material. Both the retrosynthetic analysis and the actual synthesis of 6 are shown in Figure 9.3.

C. Product Analysis. The stereoisomeric products 7 and 8 must be separated *and identified* before the necessary information can be extracted. Differentiation between stereoisomeric, trisubstituted alkenes is not easy by spectroscopy and so Roth and coworkers decided on independent synthesis of one compound, 8. In order to achieve this they made use of the reaction that they were studying, the [1,5] H migration. One might be concerned that application of the [1,5]

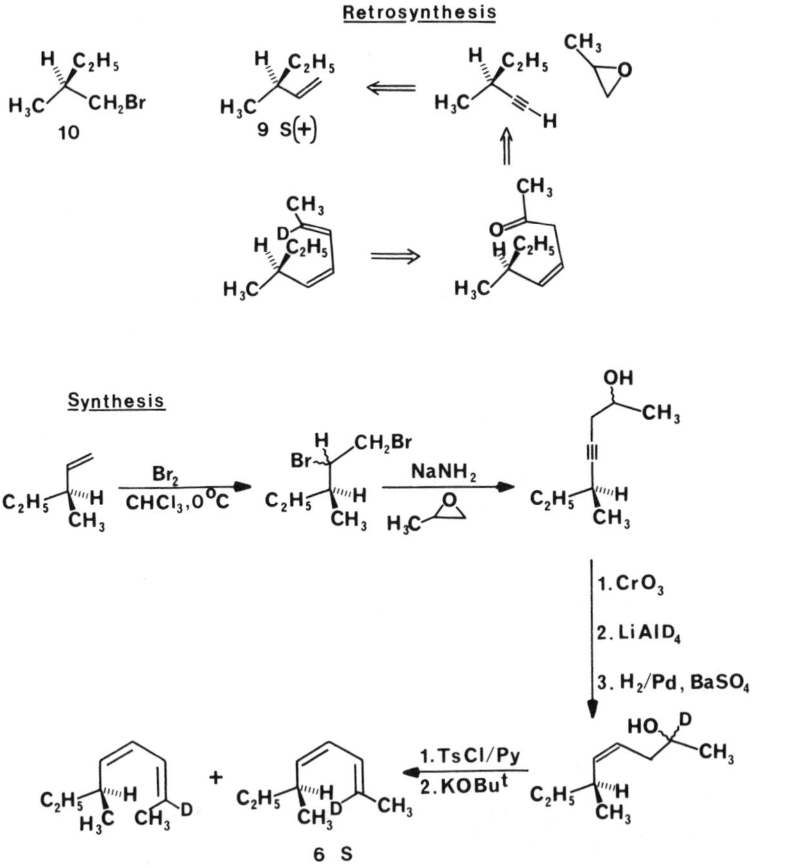

Retrosynthesis

Synthesis

FIGURE 9.3. Retrosynthetic analysis and synthetic sequence leading to the preparation of **6** with known absolute configuration.

sigmatropic shift in the product analysis hints of circular reasoning but, in fact, this use of the reaction does not require any knowledge of its stereochemistry. The reactant for this sigmatropic rearrangement was synthesized as shown on p. 184.

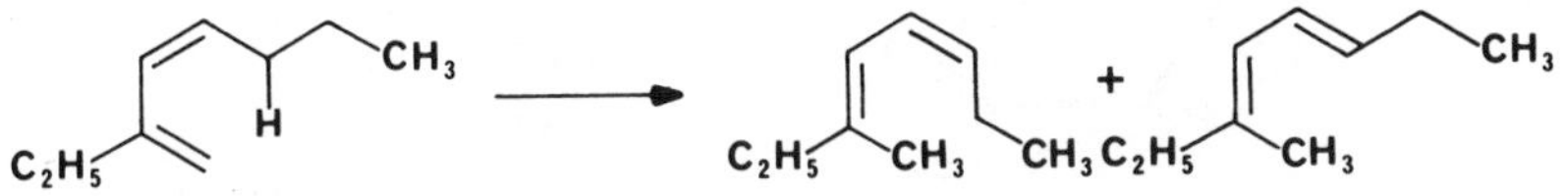

The remaining analytical problem was to relate the absolute configurations of the products to that of the reactant. This task was made somewhat easier by virtue of the fact that the absolute configuration of the reactant **6** was known. The products **7** and **8** were converted to a common alcohol by ozonolysis and reduction. This alcohol was then independently synthesized with known absolute configuration from lactic acid.

D. Experimental Results. S **6** was converted to **7** and **8**. The propanol-2-d from **7** showed a specific rotation of +0.398 ± 0.002° at 302 nm, indicating the R configuration. The alcohol from **8** had a rotation of −0.410 ± 0.002°, indicating the S configuration. These results are in accord with a suprafacial H migration, as predicted by the Woodward–Hoffmann rules.

9.3. THERMAL REARRANGEMENT OF 1,5-HEXADIENE DERIVATIVES

The [3,3] and [1,3] rearrangements of 1,5-hexadiene and its derivatives are among the most thoroughly studied reactions in organic chemistry. Yet there are still areas that remain to be clarified in the mechanisms. In fact one cannot claim complete "understanding" of the mechanisms until their potential energy surfaces have been mapped out as a function of the various geometrical coordinates of the molecule. Of course the full potential energy surface with 43 dimensions would be an immense task to explore but a restricted version in which fewer geometrical coordinates are varied would be more reasonable. One possibility would be to determine how the reaction responds to changes in the dihedral angles θ_{23} and θ_{34} (with θ_{45} fixed at $90°$). Some points in this map are known (see Figure 9.4). For example *trans*-divinylcyclobutane, with $\theta_{23} = 90°$ and $\theta_{34} = 120°$ exhibits biradical and [1,3] reactions; 1,5-hexadiene in its chair conformation has $\theta_{34} = 90°$ and $\theta_{34} = 60°$, and shows a clear preference for the [3,3] rearrangement; boat 1,5-hexadiene with $\theta_{23} = 270°$ and $\theta_{34} = 0°$ also shows the [3,3] mechanism to be lowest lying but the activation energy is greater than that for the chair.

These few points are not nearly sufficient to define the behavior of the various [3,3], [1,3], and biradical surfaces at all values of θ_{23} and θ_{34}. The study described in this section adds one more point. It concerns the rearrangement of 6-methylenebicyclo[3.2.2]oct-2-ene.

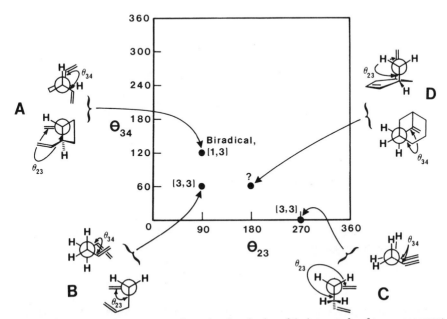

FIGURE 9.4. Some points on a three-dimensional projection of the hypersurface for rearrangement of 1,5-hexadienes. A is *trans*-divinyl cyclobutane, B is 1,5-hexadiene in a chair conformation, C is 1,5-hexadiene in a boat conformation, and D is the subject of the present study, 6-methylene-bicyclo[3.2.1]oct-2-ene.

A. The Strategy.   At the inception of this study by Berson and Janusz[8] there was already in the literature a similar investigation of the thermal rearrangement of 6-methylenebicyclo[3.2.0]hept-2-ene by Hasselmann.[9] However, in this system not all rearrangements are degenerate, and one could imagine that the exothermic [1,3] and [3,3] rearrangements might be artifically favored over the degenerate [1,3] shift.

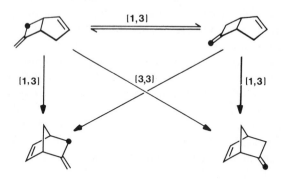

In the homologue studied by Berson and Janusz all rearrangements are degenerate as they are for the parent 1,5-hexadiene. One may thus assume that the relative favorabilities of each reaction reflect the effects of changes in the dihedral angles rather than spurious effects due to changes in strain energy.

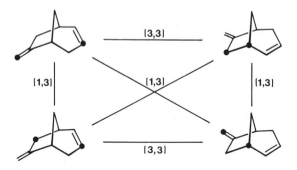

The advantages of using degenerate rearrangements are counterbalanced by an increase in the experimental difficulty of their analysis. In a degenerate rearrangement one cannot simply conduct an experiment with a labeled reactant and deduce the location of the label in the product since, by definition, the "product" is undergoing rearrangements at the same rate as the reactant. Under these circumstances it becomes necessary to carry out a kinetic analysis of the problem. In the present case the analysis is amenable to solution by the matrix techniques described in Chapter 4 provided that one ignores secondary isotope effects. We will discuss this approximation later. The full mechanistic problem is depicted in Figure 9.5.

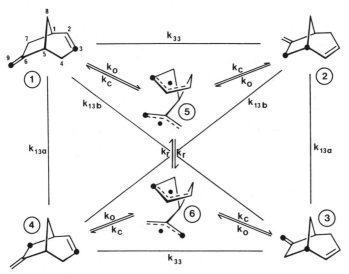

FIGURE 9.5. Pericyclic and biradical mechanisms for the degenerate rearrangement of 6-methylenebicyclo[3.2.1]oct-2-ene.

The symmetrized rate constant matrix for this problem is

$$
\mathbf{K}_s =
\left[
\begin{array}{cccc|cc}
k_t & -k_{33} & -k_{13b} & -k_{13a} & -k_{oc} & 0 \\
-k_{33} & k_t & -k_{13a} & -k_{13b} & -k_{oc} & \\
-k_{13b} & -k_{13a} & k_t & -k_{33} & 0 & -k_{oc} \\
-k_{13a} & -k_{13b} & -k_{33} & k_t & 0 & -k_{oc} \\
\hline
-k_{oc} & -k_{oc} & 0 & 0 & k_{cr} & -k_r \\
0 & 0 & -k_{oc} & -k_{oc} & -k_r & k_{cr}
\end{array}
\right]
$$

where $k_t = k_{13a} + k_{13b} + k_{33} + k_o$
$k_{oc} = \sqrt{(k_o k_c)}$
$k_{cr} = 2k_c + k_r$

After application of the steady-state approximation to the two biradical concentrations one can find the eigenvalues and eigenvector matrix. They are

$$\lambda_1 = 0$$

$$\lambda_2 = 2(k_{33} + k_{13b}) + k_o$$

$$\lambda_3 = 2(k_{33} + k_{13a}) + k_o$$

$$\lambda_4 = 2(k_{13a} + k_{13b}) + \frac{k_o k_r}{(k_c + k_r)}$$

$$\mathbf{B}_s = 0.5 \begin{bmatrix} 1 & 1 & 1 & 1 \\ 1 & -1 & -1 & 1 \\ 1 & -1 & 1 & -1 \\ 1 & 1 & -1 & -1 \end{bmatrix}$$

from which the following integrated rate equations can be found:

$$A_1 = 0.25(S_1 + S_2 e^{-\lambda_2 t} + S_3 e^{-\lambda_3 t} + S_4 e^{-\lambda_4 t})$$

$$A_2 = 0.25(S_1 - S_2 e^{-\lambda_2 t} - S_3 e^{-\lambda_3 t} + S_4 e^{-\lambda_4 t})$$

$$A_3 = 0.25(S_1 - S_2 e^{-\lambda_2 t} + S_3 e^{-\lambda_3 t} - S_4 e^{-\lambda_4 t})$$

$$A_4 = 0.25(S_1 + S_2 e^{-\lambda_2 t} - S_3 e^{-\lambda_3 t} - S_4 e^{-\lambda_4 t})$$

$$S_1 = A_1^\circ + A_2^\circ + A_3^\circ + A_4^\circ$$

$$S_2 = A_1^\circ - A_2^\circ - A_3^\circ + A_4^\circ$$

$$S_3 = A_1^\circ - A_2^\circ + A_3^\circ - A_4^\circ$$

$$S_4 = A_1^\circ + A_2^\circ - A_3^\circ - A_4^\circ$$

The next problem is to consider the possible experiments that would allow one to determine the phenomenological rate constants $\lambda_2 - \lambda_4$.

One plausible experiment would be to follow the label at C3 by NMR. Its fractional integral would be given by

$$I_3 = A_1 + A_4$$

$$= 0.5(S_1 + S_2 e^{-\lambda_2 t})$$

Thus the phenomenological rate constant for decrease in the integral of this NMR resonance would be λ_2.

Similarly,

$$I_9 = A_1 + A_3$$

$$= 0.5(S_1 + S_3 e^{-\lambda_3 t})$$

The rate constant for decrease in the integral of the NMR resonance for the label at C9 would be λ_3.

Finding λ_4 is not so easy. Measuring racemization of an optically active reactant is no use because

$$\alpha_{obs} = [\alpha](A_1 + A_4 - A_2 - A_3)$$

$$= [\alpha]S_2 e^{-\lambda_2 t}$$

Hence, $k_\alpha = \lambda_2$, which we already have.

In order to determine λ_4 it will be necessary to re-resolve the reaction mixture and determine the label locations in the separate enantiomers.

B. The Synthesis. Retrosynthetic analysis of the problem must take into account the need for an optical resolution and must consider methods for introducing the isotopic labels (whose exact natures are not yet specified). The label at C9 is easy because it can be introduced by a Wittig reaction. The label at C3 can be introduced by means of a cyclopropanation followed by a ring expansion. Preparation of norbornenone from *endo*-norbornene carboxylic acid would allow the optical resolution.

The choice of isotopic labels is influenced by questions of economics and ease of synthesis as well as by the desire to minimize equilibrium isotope effects. The first two would lead to a preference for deuterium whereas the third would lead to a preference for ^{13}C. As it turned out a compromise was possible. It was found that the equilibrium constant for

was within experimental error of 1.00, in other words the equilibrium isotope effect was negligible. On the other hand for the equilibrium constant was 1.04 and so ^{13}C was used at this site. The final synthesis proceeded as shown in Figure 9.6.

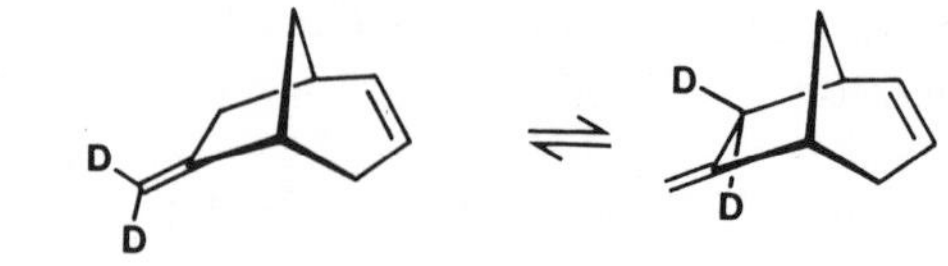

C. Product Analysis.   The migration of the isotopic labels was followed by ^1H and ^{13}C NMR. Reresolution of the reaction mixture was achieved by the sequence shown below. The acid was resolved with cinchonidine.

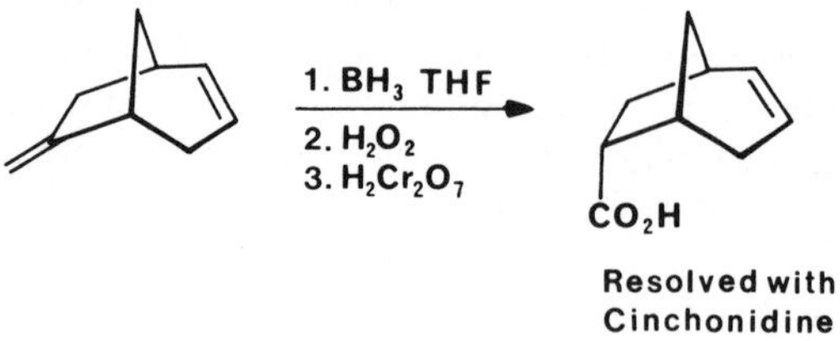

D. Experimental Results.   The three independent eigenvalues were found to be

$$\lambda_2 = 2(k_{33} + k_{13b}) + k_o = (1.95 \pm 0.06) \times 10^{-4} \text{ sec}^{-1}$$

$$\lambda_3 = 2(k_{33} + k_{13a}) + k_o = (1.82 \pm 0.09) \times 10^{-4} \text{ sec}^{-1}$$

$$\lambda_4 = 2(k_{13a} + k_{13b}) + \frac{k_o k_r}{(k_c + k_r)} = (0.42 \pm 0.09) \times 10^{-4} \text{ sec}^{-1}$$

all at 344° C. Actually these equalities are not strictly correct since there could be small secondary kinetic isotope effects of unknown magnitude. The choice of labels addressed only the problem of equilibrium isotope effects. Berson and Janusz handled this problem by performing numerical integrations to refine the fit of the mechanistic rate constants to the observed concentration changes. The corrections were small and will be ignored here.

The first thing that is obvious from inspection of the three equations above is that the problem is underdetermined and that assignment of unambiguous values to the mechanistic rate constants is not possible. However some conclusions can be drawn. Let us begin by assuming that, in accord with Occam's razor, we are looking for the simplest hypothesis that will fit the facts. The simplest hypothesis would be that there is a single process that is responsible for formation of all

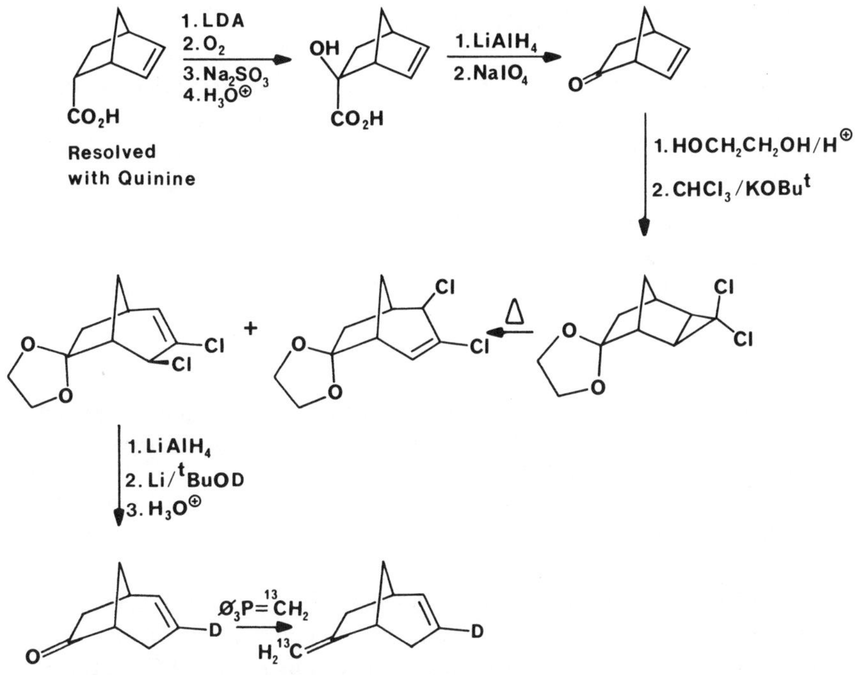

FIGURE 9.6. Synthesis of optically active, labeled 6-methylenebicyclo[3.2.1]oct-2-ene.

products. If this were a [1,3] sigmatropic shift then one would be able to set k_{33} and $k_o = 0$, leading to the prediction that $\lambda_4 = \lambda_2 + \lambda_3$. This is clearly incorrect.

An exclusive [3,3] sigmatropic shift would mean that $k_{13a} = k_{13b} = k_o = 0$, leading to the prediction that $\lambda_4 = 0$. This, too, is incorrect.

An exclusive biradical process would mean that $k_{13a} = k_{13b} = k_{33} = 0$, leading to the prediction that $\lambda_2 = \lambda_3$. This is possible within the experimental error. The data can be fit within experimental error if $k_o = 1.89 \times 10^{-4} \sec^{-1}$ and $k_c/k_r = 3.5$. This explanation would mean that the biradical existed in a very shallow potential well. In fact if $\Delta G^{\ddagger}_{rotation}$ has a typical value of ≈ 2 kcal/mol then $\Delta G^{\ddagger}_{closure}$ can be only ≈ 0.5 kcal/mol at $344°$C.

If one insisted that the biradical be a true intermediate sitting in a deep potential well such that $k_r \gg k_c$ then $k_r/(k_c + k_r) \simeq 1$ and an exclusive biradical pathway would predict $\lambda_2 = \lambda_3 = \lambda_4$, which is incorrect. With this definition of a biradical the simplest mechanism that would fit would be a mixture of [3,3] and biradical processes such that $k_{33} = 0.73 \times 10^{-4} \sec^{-1}$ and $k_o = 0.42 \times 10^{-4}$. This would correspond to a difference in activation free energy of 0.7 kcal/mol between the two processes.

It thus appears that the dihedral angles $\theta_{23} = 180°$ and $\theta_{34} = 60°$ have brought the biradical surface at least very close to and possibly below the surface for the pericyclic [3,3] rearrangement.

9.4.   THE "ENE" REACTION OF $^1\Delta$ OXYGEN

Molecular oxygen in its first excited single state ($^1\Delta$) will react with most alkenes that bear an allylic hydrogen to give a hydroperoxide, in a reaction whose gross features look similar to those of the "ene" reaction.[10] A number of mechanisms

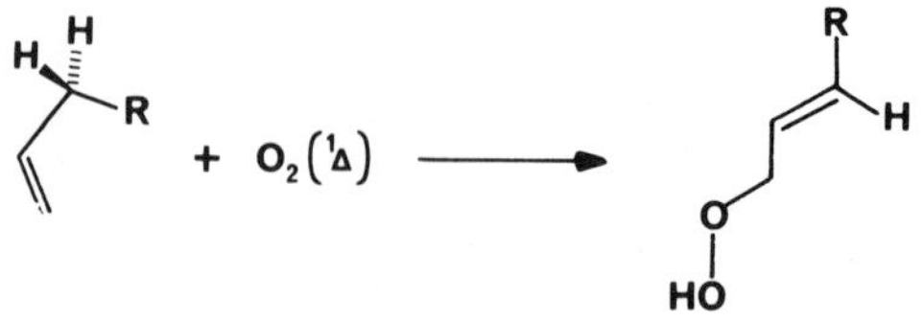

have been considered for this process. The three most common are depicted in Figure 9.7. The following experiments designed to distinguish between them were carried out by Stephenson and coworkers.[11,12]

9.4.1.   Chirality Studies

A.   The Strategy.   The first step was to distinguish the biradical mechanism B from the pericyclic ene mechanism A and the "perepoxide" mechanism C.

In order to do this one must notice that in mechanisms A and C the forming C—O bond and the breaking C—H bond are suprafacially related. In mechanism B the intermediate biradical is assumed to undergo free rotation, meaning that formation of the C—O bond and cleavage of the C—H bond do not have to occur on the same face of the alkene.

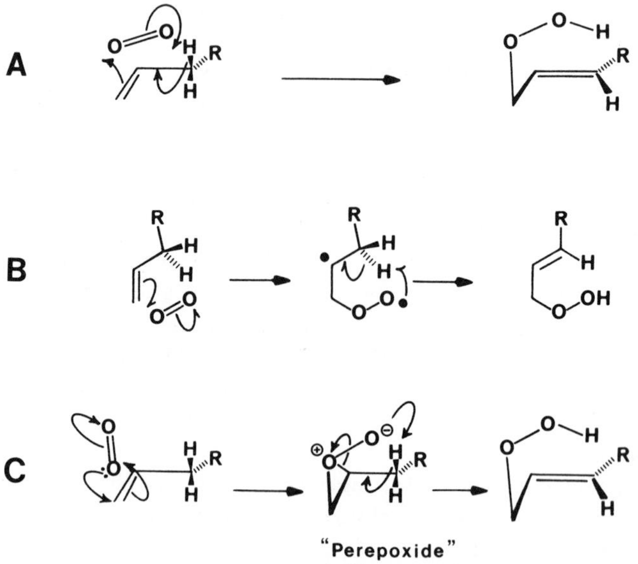

FIGURE 9.7.   Mechanisms for the "ene" reaction of $O_2(^1\Delta)$ with alkenes.

The strategy used for testing this was quite similar to the one employed by Roth in the study of the [1,5] H migration (Section 9.2). An alkene that was optically active at the allylic carbon by virtue of an isotopic substitution, and of known configuration about the double bond was used as substrate. The product analysis entailed relating the configuration of the new chiral center to that of the reactant and to the presence of H or D on the new double bond.

The choice of R_1, R_2, and R_3 was influenced by these analytical requirements. Thus alcohol **11** of known absolute configuration had been reported. This meant

11 S
$[\alpha]_D^{20} = +28.0°$

that preparation of the alkene **12** with known absolute configuration would allow determination of all of the necessary information, provided that it underwent the desired reaction with singlet oxygen. A control experiment with the unlabeled analog showed that it did.

94 %

B.   The Synthesis.   A plausible retrosynthesis takes one from **12** with the R absolute configuration to commercially available S lactic acid:

The details of the transformations that allowed the forward synthesis were not reported.[11]

C.   Product Analysis.   It is not sufficient in this experiment merely to measure total optical activity and total deuterium content of the products. In order to deduce the necessary mechanistic information one must determine the deuterium content of individual enantiomers.

This was achieved by the use of Mosher's reagent (**13**), which converts enantiomeric alcohols to disastereomeric esters. These usually exhibit resolvable NMR resonances. The principle is much the same as that for optically active lanthanide shift reagents.

13

The peaks in the ^{13}C—NMR spectrum of the mixture of esters could be identified by preparing the ester from the known S(+) alcohol. The location of the deuterium could be determined by the presence of a triplet resonance (due to ^{13}C—D coupling) in the ^{13}C{^{1}H} NMR.

D.   Experimental Results.   The total deuterium content of the allylic alcohol was determined by mass spectrometry and by ^{1}H NMR. The result was D/H = 1.04. The total optical activity of the allylic alcohol was determined by optical rotation and by ^{19}F NMR on the esters of Mosher's reagent. The result was S/R = 1.05. Finally, the ^{13}C NMR of the Mosher's esters showed only a singlet resonance for C2 in the ester of the R alcohol and only a triplet in the ester of the S alcohol. The conclusion is that the product is approximately 50% R(H) and 50% S(D) and that no R(D) or S(H) products can be detected.

This result is consistent with mechanisms A or C but inconsistent with mechanism B if one assumes a freely rotating biradical. The apparent lack of a primary isotope effect $(D/H = 1.04)$ would be consistent with mechanism C if formation of the "perepoxide" were rate determining, but would be more difficult to explain for mechanism A where a C—H bond is being broken in the rate-determining step.

In order to make a more definitive distinction between mechanism A and C one could hope to make use of the technique in which intermediates are detected by comparing inter- and intramolecular isotope effects (Section 5.4). Unfortunately rate measurements are difficult in the present case because the singlet oxygen is generated photochemically. Stephenson and coworkers circumvented this problem by comparing two different intramolecular isotope effects, thereby allowing all measurements to be made on the final product ratio.

9.4.2. Isotope Effect Studies

A. The Strategy. The special feature of the present reaction that makes it possible to use two intramolecular isotope effects instead of one inter- and one intra- is the fact that the geometry of the hypothetical perepoxide in mechanism C allows competition for H removal only between *cis*-related vicinal alkyl groups. In mechanism A (or, for that matter, mechanism B) all four alkyl groups in a tetrasubstituted alkene are in competition. Thus if one compared intramolecular isotope effects for 2,3-dimethyl-2-butene-d_6 isomers one could hope to make a distinction between the two mechanisms. The explicit predictions for mechanisms A and C using *cis*-labeled and *trans*-labeled substrate are shown in Figure 9.8.

B. The Synthesis. Preparation of the requisite labeled alkenes was achieved by reduction of the corresponding esters:

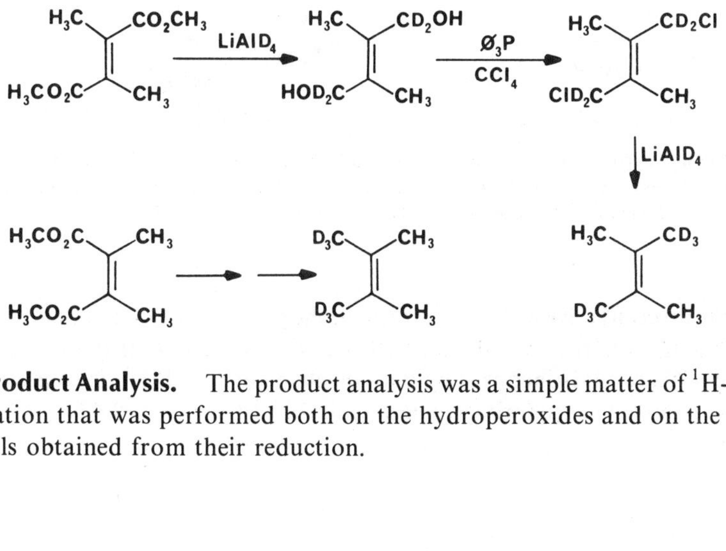

C. Product Analysis. The product analysis was a simple matter of ¹H-NMR integration that was performed both on the hydroperoxides and on the allylic alcohols obtained from their reduction.

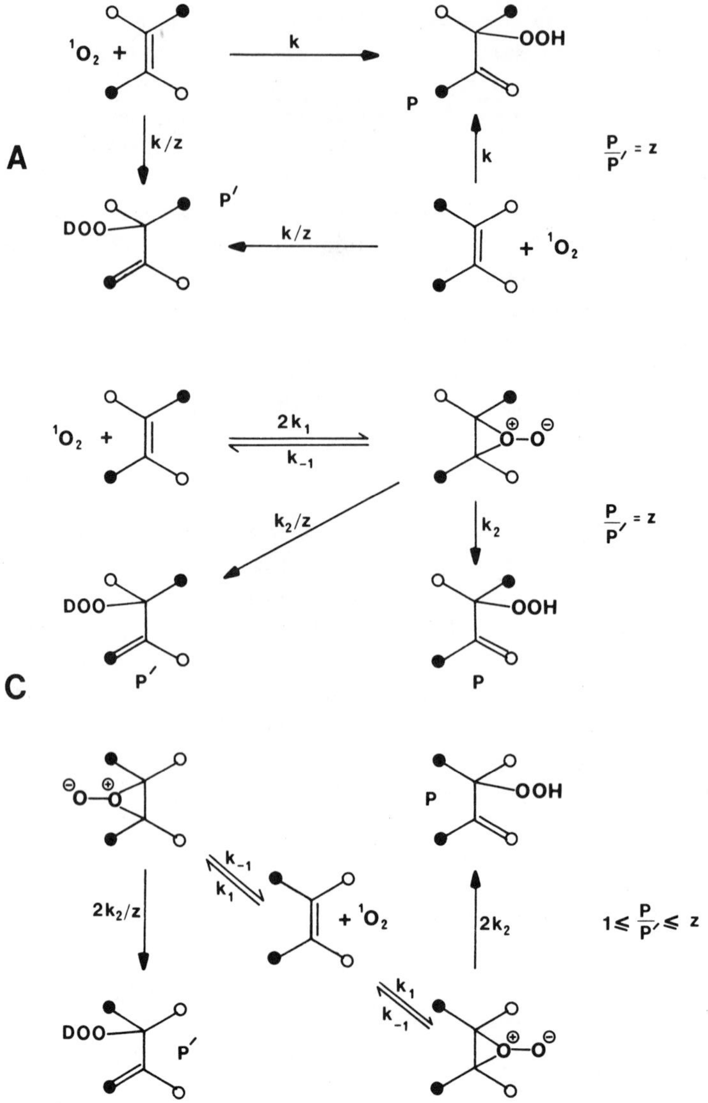

FIGURE 9.8. Predictions for mechanisms A and C using stereoisomers of tetramethyl ethylene-d_6. The open circles are CH_3 or CH_2. The filled circles are CD_3 or CD_2. z is the intrinsic primary isotope effect that accompanies cleavage of a C—D rather than a C—H bond.

D. Experimental Results.   The ratio of the products P : P′ (see Figure 9.8) was 1.40 ± 0.02 from the *trans*-labeled alkene and 1.07 ± 0.03 from the *cis*-labeled alkene. This result is inconsistent with mechanism A or B but is consistent with mechanism C. The quantitative data would be fit by $k_{-1} = 0.42\, k_2$ (see Figure 9.8).

9.5. THE STEREOCHEMISTRY OF BUTADIENE DIMERIZATION

We have seen several times the controversy surrounding the mechanism of the Diels–Alder reaction. The argument about the intermediacy of a biradical has been contentious, and nowhere more so than in the dimerization of butadiene:

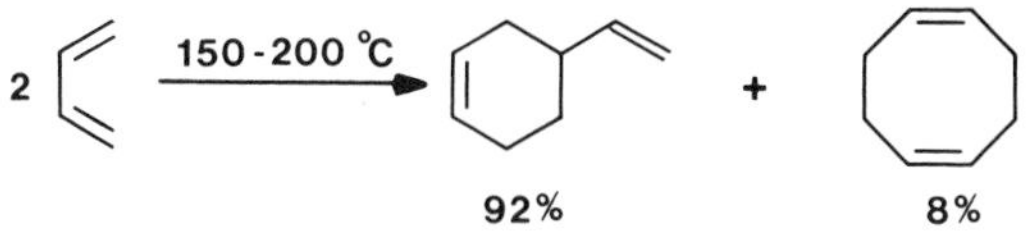

The special attraction of invoking a biradical in this case is that it could be imagined to link a number of C_8H_{12} rearrangements:

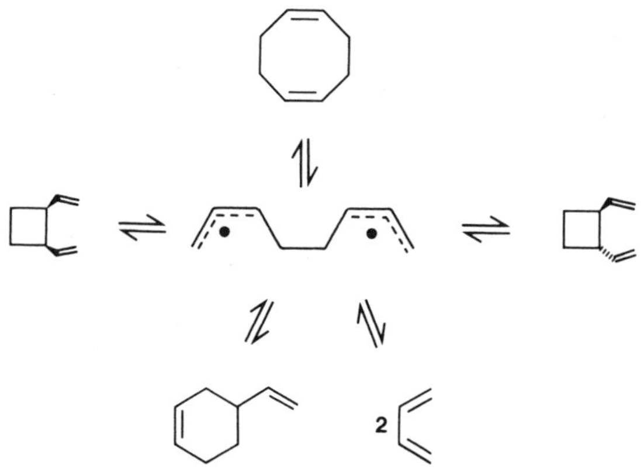

It thus appears that butadiene dimerization is an especially important place to look for the stereochemical scrambling that might be indicative of a biradical intermediate.

A. The Strategy. The experiment obviously involves stereospecific labeling, but the product analysis is a formidable challenge. If one uses deuterium labels alone it is not possible (at least by present analytical techniques) to determine the complete stereochemistry of the reaction. Use of a combination of alkyl substituents and deuterium labels allows a complete stereochemical analysis but at the price of a loss in "rigor" of the experiment (if one is interested in the intrinsic behavior of the butadiene system then chemical modification by introduction of alkyl substituents is undesirable). As it happens both approaches have been used, the first by Stephenson[13] and the second by Berson.[14] The reader can decide which is more appealing. We will begin with the butadiene dimerization.

B. The Synthesis. The three possible isomers of 1,3-butadiene-1,4-d_2 were prepared by zinc-copper couple reduction of the corresponding chlorides in D_2O/dioxane. They were analyzed by Raman spectroscopy. Dienes reisolated from an incomplete reaction showed no stereochemical scrambling.

C. Product Analysis. ^{1}H-NMR analysis of the vinylcyclohexene label isomers from dimerization of *cis,cis*-1,3-butadiene-1,4-d_2 showed that the vinyl groups had only *cis*-deuterium labels:

$$J_{HH} = 11,17 \text{ Hz} \qquad\qquad J_{HH} = 11 \text{ Hz}$$

Further analysis was hampered by the mutual proximity of the NMR resonances. This problem was alleviated by derivatizing the vinylcyclohexene and then using a lanthanide NMR shift reagent. The resonances in the bicyclic ether were

assigned by decoupling experiments. From these assignments it was possible to predict intensities of the various lines for the *exo* and *endo* stereochemistries of a $_\pi 2_s + _\pi 4_s$ pericyclic reaction and for a reaction involving a freely rotating biradical.

	Integrated Intensities		
	H_a	H_b	H_c
Endo $_\pi 2_s + _\pi 4_s$	0.0	1.0	1.0
Exo $_\pi 2_s + _\pi 4_s$	1.0	1.0	0.0
Freely rotating biradical	—	0.5	—
Observed	0.5	0.9	0.5

The authors concluded that the reaction was mostly $_\pi 2_s + _\pi 4_s$ with about a $1:1$ *exo*:*endo* ratio. Note, however, that $1:1$ *exo*:*endo* $_\pi 4_s$ is indistinguishable from $1:1$ *exo*:*endo* $_\pi 4_a$ because mixtures I and II below would be indistinguishable by this analysis technique. The approximately 10% scrambling at H_b could be

consistent with 20% biradical mechanism but the authors considered this unlikely since no more than 8% of the total reaction product could have come from a *trans,trans*-biradical (which would give divinyl cyclobutanes and, eventually the cyclooctadiene) and so it would be surprising if 20% had come from a *cis,trans*- or *cis,cis*-biradical since one might have guessed that these would be higher energy configurations. The authors preferred to interpret the results in terms of a 10% contribution from $_\pi 2_a$ mode of cycloaddition (whether $_\pi 2_a + _\pi 4_s$ or $_\pi 2_a + _\pi 4_a$ could not be determined).

Berson and coworkers[14] attacked the problem by using stereospecifically deuterated *trans*-piperylene. The advantage was that the diastereomeric methylpropenylcyclohexenes could be separated and identified.

B.   The Synthesis.   The labeled diene reactant was prepared from *trans*-pent-3-ene-1-yne by reductive hydroboration:

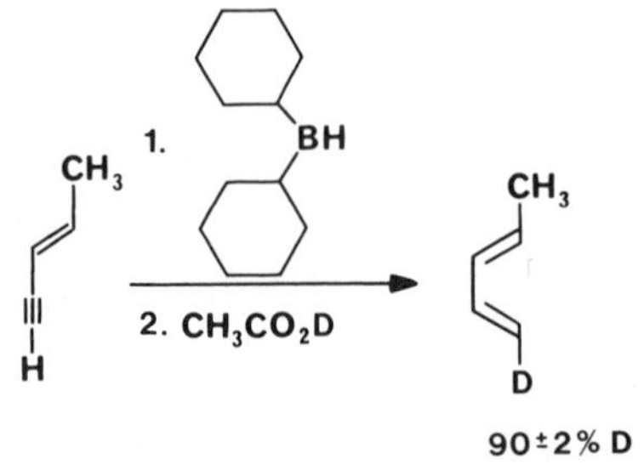

90±2% D

C.   Product Analysis.   The products of dimerization at 195°C consisted of 82% 3-methyl-4-propenycyclohexenes of which 45% was *trans* and 55% *cis*. These separated diastereomers were derivatized and treated with Eu(fod)$_3$ prior to NMR analysis. The results were:

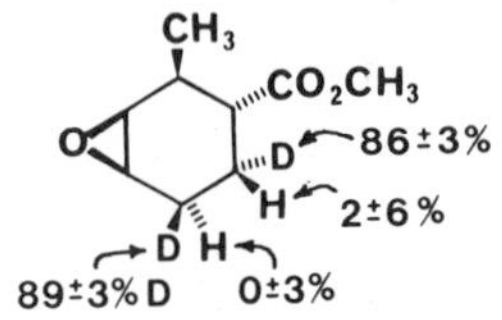

A completely stereospecific $_\pi 2_s + _\pi 4_s$ would have given:

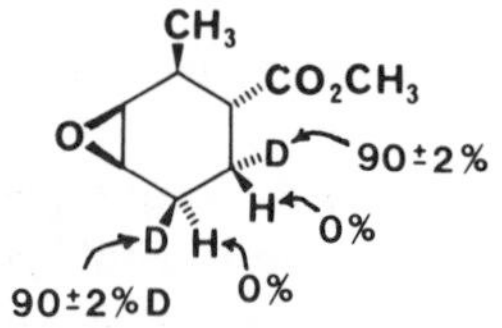

which appears to be within experimental error of the observed result. The other diastereomer of the 3-methyl-4-propenylcyclohexene gave similar results.

The difference between the Stephenson and Berson results is interesting. It might mean that, indeed, introduction of a methyl group does change the mechanism somewhat. If, on the other hand, the discrepancy is an experimental artifact then the problem almost certainly has to be in Stephenson's results since it is very hard to think of any way that the reaction could be made to appear *more* stereoselective than it really is. One possible problem might have occurred in the sodium borohydride reduction of Stephenson's epoxy-aldehydes. Since this reaction generates sodium methoxide as a byproduct one might be concerned about epimerization of the aldehyde prior to reduction. There appears to be no evidence to support or refute this speculation.

9.6. MECHANISMS OF VITAMIN B$_{12}$ REACTIONS

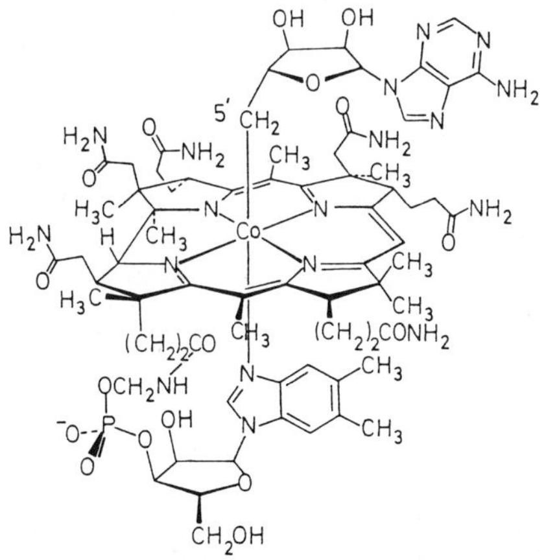

Vitamin B$_{12}$ Coenzyme

I have included this final example (i) to show that the techniques we have discussed are not limited to analysis of esoteric hydrocarbon rearrangements and (ii) to point out some of the extremely elegant mechanistic work that is now being done in the bioorganic field.

The vitamin B$_{12}$ coenzyme is involved in a number of important transformations,[15] some of which are shown below. We will concentrate on (c) and (d).

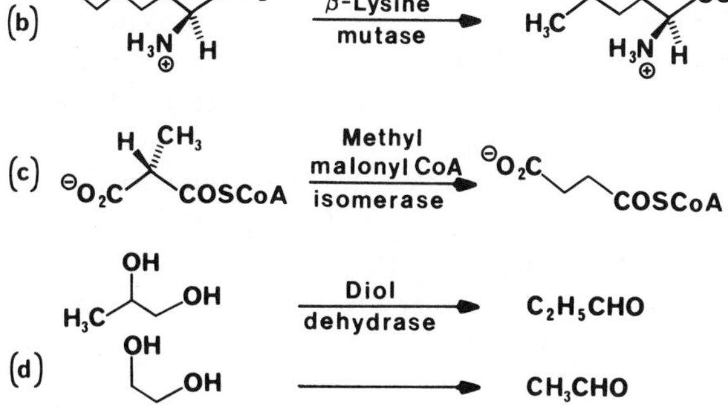

All of the transformations except (d) appear to be of the type:

$$\underset{H\quad X}{\overset{R}{\diagdown}} \rightleftharpoons \underset{X\quad H}{\overset{R}{\diagdown}}$$

In the case of (c) one could imagine that either the carboxyl group or the COSCoA group migrated. Eggerer[16] showed it to be the latter by a ^{14}C-labeling

$$\underset{COSCoA}{\overset{^{14}\bullet\quad CO_2^\ominus}{\diagdown}} \xrightarrow{B_{12}} \underset{COSCoA}{\overset{CO_2^\ominus}{\diagdown}}$$

experiment. The intramolecularity of the rearrangement was demonstrated with a double-labeling crossover experiment by Kellermeyer and Wood.[17]

$$\left\{ \begin{array}{c} ^{13}C \rightarrow \overset{CO_2^\ominus}{COSCoA} \\ \overset{CO_2^\ominus}{COSCoA} \end{array} \right\} \longrightarrow \left\{ \begin{array}{c} CoASOC \quad \overset{CO_2^\ominus}{CO_2^\ominus} \\ CoASOC \end{array} \right\}$$

The apparently anomalous (d) transformation could be made to fit into the general scheme if one considered a mechanism involving oxygen migration rather than dehydration to the enol. These two mechanisms could be distinguished by [18]O labeling:

$$\underset{^{18}O\nearrow}{\overset{H_3C\quad H}{\underset{HO\quad OH}{\diagup}}} \longrightarrow \overset{H_3C}{\underset{OH}{\diagup}} \longrightarrow C_2H_5CHO$$

$$\underset{HO\quad H}{\overset{H_3C\quad OH}{\diagup}} \longrightarrow \underset{H\quad OH}{\overset{H_3C\quad OH}{\diagup}} \longrightarrow C_2H_5CHO + C_2H_5CHO$$

Arigoni carried out this experiment in 1966.[18] Using S propanediol with the oxygen on C2 being labeled he showed that the product propionaldehyde contained no label. At first sight this would appear to rule out the oxygen migration mechanism, but that is not correct. When the R enantiomer of the same label isomer was used the product contained 100% of the label! This apparently incomprehensible result is, in fact, quite familiar to bioorganic chemists. It is caused by the participation of an enzyme in the reaction. Thus the enantiotopic hydroxyls in the hydrate that would result from oxygen migration become *diastereotopic* in the chiral environment of an enzyme active site. It is quite common for there to be a large preference for loss of one group over another apparently enantiotopically related group under these circumstances.

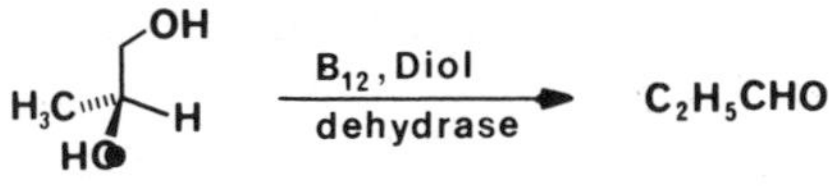

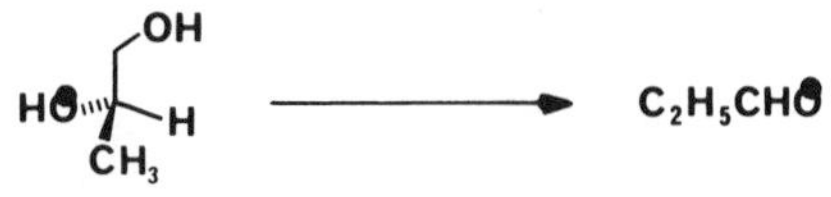

$$\text{(structure: CH}_3\text{CHOH-CHOH-H with OH)} \xrightarrow[\text{dehydrase}]{B_{12}, \text{Diol}} C_2H_5CHO$$

$$\text{(structure with OH, HO, CH}_3) \longrightarrow C_2H_5CHO$$

Arigoni's result was thus consistent with the oxygen migration mechanism and further showed that the migration was stereospecific and that the dehydration step was enzyme mediated.

Abeles[19] showed that reaction (d) involved rate determining C—H bond cleavage by observing a very large isotope effect ($k_H/k_D = 12$) for the labeled

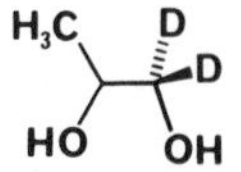

$$\underset{HO \quad OH}{\overset{H_3C \quad D}{\diagdown \; \diagup}} D$$

substrate shown below. This was much larger than the isotope effect observed for the formally related pinacol rearrangement of propanediol ($k_H/k_D = 3$), suggesting that the mechanism was different.

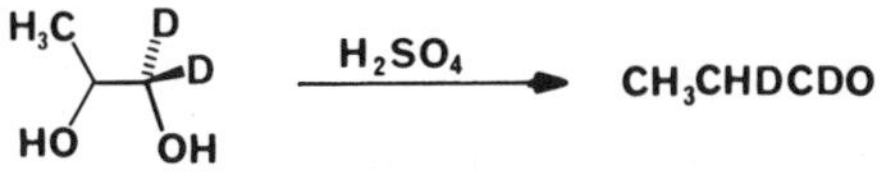

$$\underset{HO \quad OH}{\overset{H_3C \quad D}{\diagdown \; \diagup}} D \xrightarrow{H_2SO_4} CH_3CHDCDO$$

An important contribution to the mechanistic puzzle was made when it was found that the H migration in reaction (d) was intermolecular:[19]

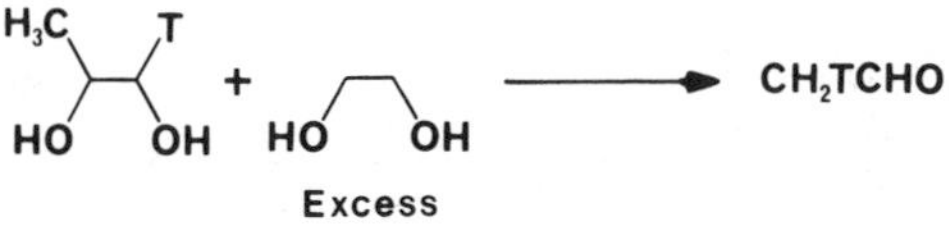

$$\underset{HO \quad OH}{\overset{H_3C \quad T}{\diagdown \; \diagup}} T \;+\; \underset{HO \quad OH}{\diagup \diagdown} \longrightarrow CH_2TCHO$$

Excess

The product showed substantial loss of tritium. The "missing" T was found at the 5' position of the adenosine unit in the vitamine B_{12} coenzyme. A similar result was obtained for reaction (c).

A further contribution came from running the reaction in an ESR spectrometer, which showed a strong signal due to Co(II).

Putting together all of the data one can write the mechanism shown in Figure 9.9. The demonstrated intramolecular and stereospecific nature of the hydroxyl migration could be consistent with this mechanism if it occurred at the active site of an enzyme. It is not so clear how the rearrangement mechanism would apply to reaction (c) although Dowd[20] has shown a similar transformation *in vitro*. The mechanism of this rearrangement is still unclear.

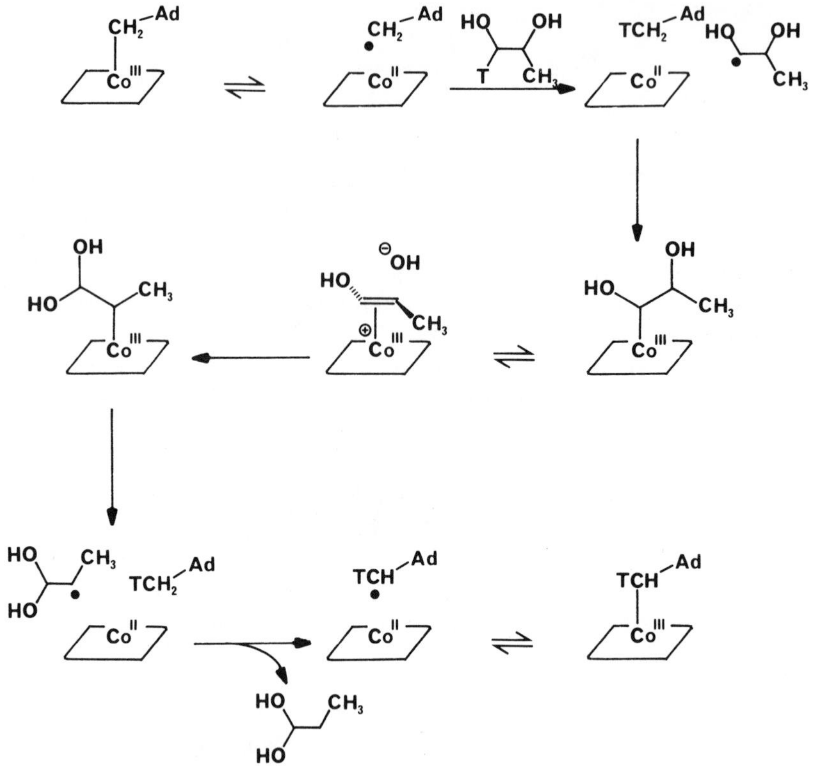

FIGURE 9.9. One possible mechanism for the vitamin B_{12} catalyzed rearrangement of propylene glycol.

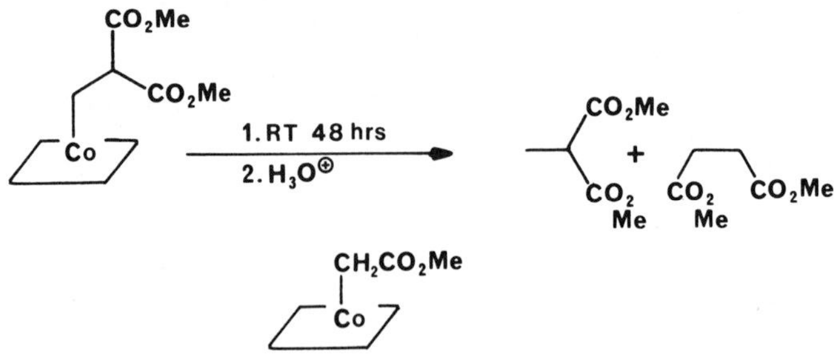

Recently Abeles and Jencks[21] have presented evidence for a π-bound enolate in the cyanide-promoted cleavage of methyl acetate from **14**.

There clearly is much work still to be done on the mechanism of action of this fascinating coenzyme and it seems reasonable to assume that at least some of the techniques described in Chapters 1–8 will contribute to unraveling the problem.

REFERENCES

1. J. F. Bunnett, M. M. Rauhut, D. Knutson. and G. E. Bussell, *J. Am. Chem. Soc.*, **76,** 5755 (1954).

2. J. F. Bunnett and M. M. Rauhut, *J. Org. Chem.*, **21,** 944 (1956).

3. D. Samuel, *Chem. Ind. (London),* 1318 (1960).

4. M. Rosenblum, *J. Am. Chem. Soc.*, **82,** 3796 (1960).

5. E. F. Ullman and E. A. Bartkus, *Chem. Ind. (London),* 93 (1962).

6. K. M. Ibne-Rasa and E. Koubek, *J. Org. Chem.*, **28,** 3240 (1963).

7. W. R. Roth, J. König, and K. Stein, *Chem. Ber.*, **103,** 426 (1970).

8. J. M. Janusz and J. A. Berson, *J. Am. Chem. Soc.*, **100,** 2237 (1978).

9. D. Hasselman, *Tetrahedron Lett.*, 3465 (1972); 3739 (1973).

10. H. M. R. Hoffmann, *Angew. Chem. Int. Ed. Engl.*, **8,** 556 (1969).

11. L. M. Stephenson, D. E. McClure, and P. K. Sysak, *J. Am. Chem. Soc.*, **95,** 7888 (1973).

12. B. Grdina, M. Orfanopoulos, and L. M. Stephenson, *J. Am. Chem. Soc.*, **101,** 3111 (1979).

13. L. M. Stephenson, R. V. Gemmer, and S. Current, *J. Am. Chem. Soc.*, **97,** 5909 (1975).

14. J. A. Berson and R. Malherbe, *J. Am. Chem. Soc.*, **97,** 5910 (1975).

15. R. H. Abeles and D. Dolphin, *Acc. Chem. Res.*, **9,** 114 (1976).

16. H. Eggerer, E. R. Stadtman, P. Overath, and F. Lynen, *Biochem. Z.*, **333,** 1 (1960).

17. R. W. Kellermeyer and H. G. Wood, *Biochemistry,* **1,** 1124 (1962).

18. J. Retey, A. Umani-Ronchi, J. Seible, and D. Arigoni, *Experientia,* **22,** 502 (1966).

19. P. A. Frey and R. H. Abeles, *J. Biol. Chem.*, **241,** 2732 (1966).

20. P. Dowd and M. Shapiro, *J. Am. Chem. Soc.*, **98,** 3724 (1976).

21. W. W. Reenstra, R. H. Abeles, and W. P. Jencks, *J. Am. Chem. Soc.*, **104,** 1016 (1982).

APPENDIX 1

LIST OF PROCEDURES IN LINEAR ALGEBRA FOR USE IN KINETIC ANALYSES OF CHAPTER 4

MULTIPLICATION OF MATRICES

$$\begin{bmatrix} a_{11} & a_{12} & a_{13} \\ a_{21} & a_{22} & a_{23} \\ a_{31} & a_{32} & a_{33} \end{bmatrix} \begin{bmatrix} b_{11} & b_{12} & b_{13} \\ b_{21} & b_{22} & b_{23} \\ b_{31} & b_{32} & b_{33} \end{bmatrix} = \begin{bmatrix} c_{11} & c_{12} & c_{13} \\ c_{21} & c_{22} & c_{23} \\ c_{31} & c_{32} & c_{33} \end{bmatrix}$$

$$\qquad\quad \mathbf{A} \qquad\qquad\qquad \mathbf{B} \qquad\qquad\qquad \mathbf{C}$$

$$c_{11} = a_{11}b_{11} + a_{12}b_{21} + a_{13}b_{31}$$

In general,

$$c_{ij} = \sum_{k=1}^{n} a_{ik}b_{kj}$$

where n is the number of columns of $\mathbf{A}$ and the number of rows of $\mathbf{B}$ (if these are not equal the multiplication is not possible).

Except under special circumstances matrices do not commute under multiplication, that is,

$$\mathbf{AB} \neq \mathbf{BA}$$

SOLUTION OF SIMULTANEOUS EQUATIONS, MATRIX INVERSION

Consider the equations

$$a_{11}x + a_{12}y + a_{13}z = c_1$$

$$a_{21}x + a_{22}y + a_{23}z = c_2$$

$$a_{31}x + a_{32}y + a_{33}z = c_3$$

In matrix notation this can be written

$$\begin{bmatrix} a_{11} & a_{12} & a_{13} \\ a_{21} & a_{22} & a_{23} \\ a_{31} & a_{32} & a_{33} \end{bmatrix} \begin{bmatrix} x \\ y \\ z \end{bmatrix} = \begin{bmatrix} c_1 \\ c_2 \\ c_3 \end{bmatrix}$$

$$\mathbf{A} \qquad \mathbf{X} = \mathbf{C}$$

Now define the inverse of $\mathbf{A}$, called $\mathbf{A}^{-1}$, such that

$$\mathbf{A}\mathbf{A}^{-1} = \mathbf{A}^{-1}\mathbf{A}$$

$$= \mathbf{I}$$

where $\mathbf{I}$ is the identity matrix, that is, a square matrix (in this case with three rows and three columns) consisting of diagonal elements of value 1 and off-diagonal elements of value 0. The properties of an identity matrix are similar to that of the number 1 in scalar arithmetic:

$$\mathbf{A}\mathbf{I} = \mathbf{I}\mathbf{A}$$

$$= \mathbf{A}$$

Hence, from the matrix form of the original equations:

$$\mathbf{A}\mathbf{X} = \mathbf{C}$$

$$\mathbf{A}^{-1}\mathbf{A}\mathbf{X} = \mathbf{A}^{-1}\mathbf{C}$$

$$\mathbf{X} = \mathbf{A}^{-1}\mathbf{C}$$

The inverse matrix is found from the formula:

$$\mathbf{A}^{-1} = \frac{(\widetilde{\text{adj }}\mathbf{A})}{\det \mathbf{A}}$$

where $(\widetilde{adj}\ \mathbf{A})$ is the transpose of the adjoint of $\mathbf{A}$ and det $\mathbf{A}$ is the determinant of $\mathbf{A}$. These terms are defined as follows:

$$\det \mathbf{A} = \begin{vmatrix} a_{11} & a_{12} & a_{13} \\ a_{21} & a_{22} & a_{23} \\ a_{31} & a_{32} & a_{33} \end{vmatrix}$$

$$\det \mathbf{A} = a_{11}\begin{vmatrix} a_{22} & a_{23} \\ a_{32} & a_{33} \end{vmatrix} - a_{12}\begin{vmatrix} a_{21} & a_{23} \\ a_{31} & a_{33} \end{vmatrix} + a_{13}\begin{vmatrix} a_{21} & a_{22} \\ a_{31} & a_{32} \end{vmatrix}$$

$$\begin{vmatrix} a_{22} & a_{23} \\ a_{32} & a_{33} \end{vmatrix} = a_{22}a_{33} - a_{23}a_{32}$$

Among the useful properties of determinants are the following:

1. Interchanging two rows or two columns changes the sign of the determinant.
2. A determinant with two identical rows or columns has a value of 0.
3. Addition of a multiple of one row to another row leaves the value of the determinant unchanged.
4. A determinant in which the upper right triangle or lower left triangle of elements is zero has a value equal to the product of the diagonal elements.
5. Multiplication of a determinant by a scalar is achieved by multiplying the elements in just one row or one column by that scalar.

The adjoint of a square matrix is defined as the matrix of cofactors where the cofactor of element a_{ij} has a value equal to $(-1)^{(i+j)}$ times the determinant of the elements remaining when the row i and column j are crossed out. Thus for the matrix

$$\mathbf{A} = \begin{bmatrix} a_{11} & a_{12} & a_{13} \\ a_{21} & a_{22} & a_{23} \\ a_{31} & a_{32} & a_{33} \end{bmatrix}$$

$$\alpha_{12} = -\begin{vmatrix} a_{21} & a_{23} \\ a_{31} & a_{33} \end{vmatrix}$$

$$\text{adj } \mathbf{A} = \begin{bmatrix} \alpha_{11} & \alpha_{12} & \alpha_{13} \\ \alpha_{21} & \alpha_{22} & \alpha_{23} \\ \alpha_{31} & \alpha_{32} & \alpha_{33} \end{bmatrix}$$

The transpose of a square matrix is obtained by interconverting the locations of elements a_{ij} and a_{ji}.

Thus the inverse of $\mathbf{A}$ can be written:

$$\mathbf{A}^{-1} = \frac{\begin{bmatrix} \alpha_{11} & \alpha_{21} & \alpha_{31} \\ \alpha_{12} & \alpha_{22} & \alpha_{32} \\ \alpha_{13} & \alpha_{23} & \alpha_{33} \end{bmatrix}}{\begin{vmatrix} a_{11} & a_{12} & a_{13} \\ a_{21} & a_{22} & a_{23} \\ a_{31} & a_{32} & a_{33} \end{vmatrix}}$$

EIGENVALUES AND EIGENVECTORS OF A SQUARE MATRIX

An eigenvalue problem is one that takes the form:

$$\hat{o}f = f\lambda$$

where $\hat{o}$ is an operator, f is some function, called an eigenfunction of $\hat{o}$, and λ is a scalar, called the eigenvalue of $\hat{o}$ and f. For matrices the eigenvalue problem takes the form:

$$\begin{bmatrix} a_{11} & a_{12} & a_{13} \\ a_{21} & a_{22} & a_{23} \\ a_{31} & a_{32} & a_{33} \end{bmatrix} \begin{bmatrix} b_1 \\ b_2 \\ b_3 \end{bmatrix} = \begin{bmatrix} b_1 \\ b_2 \\ b_3 \end{bmatrix} \lambda$$

The column vector $\mathbf{B}$ is called an eigenvector. An $n \times n$ matrix has n eigenvectors and n associated eigenvalues. The complete eigenvalue problem for a 3×3 matrix can thus be written:

$$\begin{bmatrix} a_{11} & a_{12} & a_{13} \\ a_{21} & a_{22} & a_{23} \\ a_{31} & a_{32} & a_{33} \end{bmatrix} \begin{bmatrix} b_{11} & b_{12} & b_{13} \\ b_{21} & b_{22} & b_{23} \\ b_{31} & b_{32} & b_{33} \end{bmatrix} = \begin{bmatrix} b_{11} & b_{12} & b_{13} \\ b_{21} & b_{22} & b_{23} \\ b_{31} & b_{32} & b_{33} \end{bmatrix} \begin{bmatrix} \lambda_1 & 0 & 0 \\ 0 & \lambda_2 & 0 \\ 0 & 0 & \lambda_3 \end{bmatrix}$$

or

$$\mathbf{AB} = \mathbf{B\Lambda}$$

$\mathbf{B}$ is a matrix of eigenvectors in which each column is one eigenvector of $\mathbf{A}$. $\mathbf{\Lambda}$ is a diagonal matrix of eigenvalues. Note that λ_i is the eigenvalue that corresponds to the eigenvector in column i. The eigenvalue problem is often written in the form:

$$\Lambda = \mathbf{B}^{-1}\mathbf{AB}$$

When $\mathbf{A}$ is a matrix whose elements are numerical constants, there are computer algorithms that will find $\mathbf{B}$, and Λ. However, when $\mathbf{A}$ contains algebraic unknowns, as is usually the case in kinetics problems (where $\mathbf{A} \equiv \mathbf{K}$, the rate constant matrix) the eigenvalues and eigenvectors must be found by hand. The following is a recipe that allows such evaluations.

EIGENVALUES

Define

$$\det \mathbf{A}' = \begin{vmatrix} a_{11}-\lambda & a_{12} & a_{13} \\ a_{21} & a_{22}-\lambda & a_{23} \\ a_{31} & a_{32} & a_{33}-\lambda \end{vmatrix} = 0$$

Expanding $\det \mathbf{A}'$ gives a polynomial in λ, whose roots are the eigenvalues of $\mathbf{A}$.

EIGENVECTORS

Evaluate the cofactors of $\det \mathbf{A}'$ along one row (usually the first). The cofactors α_{11}, α_{12}, and α_{13} will themselves be polynomials in λ. The eigenvector for a particular eigenvalue is found by substituting that eigenvalue for λ in these polynomials and then normalizing by dividing throughout by $\{\Sigma_j \, \alpha_{ij}^2\}^{1/2}$, where i defines the row along which the cofactors were evaluated. If $\mathbf{A}$ is a symmetric matrix (i.e., $a_{ij} = a_{ji}$) then the inverse of the eigenvector matrix is simply equal to its transpose:

$$\Lambda = \widetilde{\mathbf{B}}\mathbf{AB} \qquad \text{for symmetric } \mathbf{A}$$

Given some arbitrary matrix $\mathbf{D}$ ($\neq \mathbf{B}$)

$$\mathbf{G} = \mathbf{D}^{-1}\mathbf{AD}$$

$\mathbf{G}$ and $\mathbf{A}$ are called similar matrices. They have identical eigenvalues but different eigenvectors.

If $\mathbf{B}$ is the matrix of eigenvectors for $\mathbf{A}$ then it is also the matrix of eigenvectors for $\mathbf{A} + \mathbf{M}$, where $\mathbf{M}$ is a diagonal matrix equal to some scalar, m, times the identity. If the eigenvalues of $\mathbf{A}$ are $\lambda_1, \lambda_2, \ldots \lambda_n$, then the eigenvalues of $\mathbf{A} + \mathbf{M}$ will be $\lambda_1 + m, \lambda_2 + m, \ldots \lambda_n + m$.

SPECIAL PROPERTIES OF DIAGONAL MATRICES

If

$$\Lambda = \begin{bmatrix} \lambda_1 & 0 & 0 \\ 0 & \lambda_2 & 0 \\ 0 & 0 & \lambda_3 \end{bmatrix}$$

Then

$$\Lambda^m = \begin{bmatrix} \lambda_1^m & 0 & 0 \\ 0 & \lambda_2^m & 0 \\ 0 & 0 & \lambda_3^m \end{bmatrix}$$

and

$$e^\Lambda = \begin{bmatrix} e^{\lambda_1} & 0 & 0 \\ 0 & e^{\lambda_2} & 0 \\ 0 & 0 & e^{\lambda_3} \end{bmatrix}$$

APPENDIX 2

THE PRINCIPLE OF DETAILED BALANCE AND THE SYMMETRIZATION OF RATE CONSTANT MATRICES

As noted in Section 4.5.1, evaluation of the integrated rate equations for a system of reversible unimolecular reactions is much simpler if the rate constant matrix, **K**, can first be symmetrized. The procedure that allows this to be done is based on the principle of detailed balance.

For two species at equilibrium

$$1 \underset{k_{21}}{\overset{k_{12}}{\rightleftharpoons}} 2$$

$$A_2^\infty k_{21} = A_1^\infty k_{12}$$

where A_i^∞ is the equilibrium concentration of species i. For three species in equilibrium

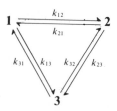

$$A_1^\infty k_{12} = A_2^\infty k_{21}$$

$$A_2^\infty k_{23} = A_3^\infty k_{32}$$

$$A_3^\infty k_{31} = A_1^\infty k_{13}$$

Hence,

$$k_{12}k_{23}k_{31} = k_{21}k_{13}k_{32}$$

In words, the product of the rate constants going around one direction of a cycle (in a complex scheme this can be any cycle) must equal the product going around in the opposite direction. This restriction represents a limitation on the values of any mechanistic rate constants that one can propose.

Turning now to the problem of matrix symmetrization, let us define

$$
S^{1/2} = \begin{bmatrix}
\sqrt{A_1^\infty} & 0 & \cdots\cdots & 0 \\
0 & \sqrt{A_2^\infty} & \cdots & 0 \\
\vdots & \vdots & & \vdots \\
0 & 0 & & \sqrt{A_n^\infty}
\end{bmatrix}
$$

also

$$
S^{-1/2} = \begin{bmatrix}
(\sqrt{A_1^\infty})^{-1} & 0 & \cdots\cdots\cdots & 0 \\
0 & (\sqrt{A_2^\infty})^{-1} & \cdots\cdots & 0 \\
\vdots & \vdots & & \vdots \\
0 & 0 & & (\sqrt{A_n^\infty})^{-1}
\end{bmatrix}
$$

and

$$\mathbf{K}_s = \mathbf{S}^{-1/2}\mathbf{K}\mathbf{S}^{1/2}$$

$$
= \begin{bmatrix}
K_{11} & L_{12} & \cdots\cdots & L_{1n} \\
L_{21} & K_{22} & \cdots\cdots & L_{2n} \\
\vdots & \vdots & & \vdots \\
L_{n1} & L_{n2} & & K_{nn}
\end{bmatrix}
$$

where

$$K_{ii} = \sum_j k_{ij}$$

and

$$L_{ij} = -k_{ji}\sqrt{(A_j^\infty / A_i^\infty)}$$

But, by the principle of detailed balance

$$k_{ij}A_i^\infty = k_{ji}A_j^\infty$$

Hence,

$$L_{ij} = L_{ji}$$

$$= -\sqrt{(k_{ij}k_{ji})}$$

Thus the matrix $\mathbf{K}_s$ is symmetric with off-diagonal elements equal to $-\sqrt{(k_{ij}k_{ji})}$.

APPENDIX 3

EIGENVALUES, EIGENVECTORS, AND SYMMETRIZATION MATRICES FOR SOME GENERALIZED KINETIC SCHEMES

In all of these solutions $\lambda_1 = 0$.

TWO-COMPONENT SYSTEM

$$1 \underset{a}{\overset{na}{\rightleftharpoons}} 2$$

$$\mathbf{K}_s = \begin{bmatrix} na & -a\sqrt{n} \\ -a\sqrt{n} & a \end{bmatrix}$$

$$\mathbf{B}_s = \begin{bmatrix} 1/p & -\sqrt{n}/p \\ \sqrt{n}/p & 1/p \end{bmatrix} \quad p = \sqrt{(n+1)}$$

$$\mathbf{S}^{-1/2} = \begin{bmatrix} p & 0 \\ 0 & p/\sqrt{n} \end{bmatrix}$$

$$\lambda_2 = (n+1)a$$

THREE-COMPONENT SYSTEM

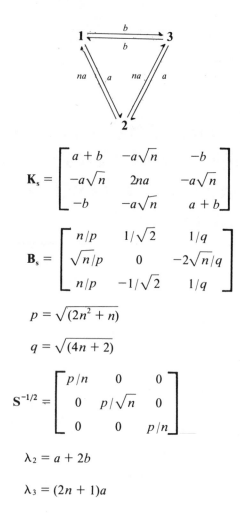

$$\mathbf{K}_s = \begin{bmatrix} a+b & -a\sqrt{n} & -b \\ -a\sqrt{n} & 2na & -a\sqrt{n} \\ -b & -a\sqrt{n} & a+b \end{bmatrix}$$

$$\mathbf{B}_s = \begin{bmatrix} n/p & 1/\sqrt{2} & 1/q \\ \sqrt{n}/p & 0 & -2\sqrt{n}/q \\ n/p & -1/\sqrt{2} & 1/q \end{bmatrix}$$

$$p = \sqrt{(2n^2 + n)}$$

$$q = \sqrt{(4n + 2)}$$

$$\mathbf{S}^{-1/2} = \begin{bmatrix} p/n & 0 & 0 \\ 0 & p/\sqrt{n} & 0 \\ 0 & 0 & p/n \end{bmatrix}$$

$$\lambda_2 = a + 2b$$

$$\lambda_3 = (2n + 1)a$$

FOUR-COMPONENT SYSTEM: CASE I

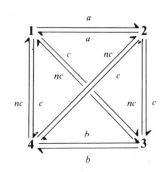

$$K_s = \begin{bmatrix} a + 2c & -a & -c\sqrt{n} & -c\sqrt{n} \\ -a & a + 2c & -c\sqrt{n} & -c\sqrt{n} \\ -c\sqrt{n} & -c\sqrt{n} & b + 2nc & -b \\ -c\sqrt{n} & -c\sqrt{n} & -b & b + 2nc \end{bmatrix}$$

$$B_s = \begin{bmatrix} \sqrt{n}/p & 1/\sqrt{2} & 0 & 1/p \\ \sqrt{n}/p & -1/\sqrt{2} & 0 & 1/p \\ 1/p & 0 & 1/\sqrt{2} & -\sqrt{n}/p \\ 1/p & 0 & -1/\sqrt{2} & -\sqrt{n}/p \end{bmatrix}$$

$$S^{-1/2} = \begin{bmatrix} p/\sqrt{n} & 0 & 0 & 0 \\ 0 & p/\sqrt{n} & 0 & 0 \\ 0 & 0 & p & 0 \\ 0 & 0 & 0 & p \end{bmatrix}$$

$$p = \sqrt{(2n + 2)}$$

$$\lambda_2 = 2a + 2c$$

$$\lambda_3 = 2b + 2nc$$

$$\lambda_4 = 2(n + 1)c$$

FOUR-COMPONENT SYSTEM: CASE II*

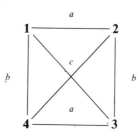

*All equilibrium constants are 1.

$$\mathbf{K}_s = \begin{bmatrix} a+b+c & -a & -c & -b \\ -a & a+b+c & -b & -c \\ -c & -b & a+b+c & -a \\ -b & -c & -a & a+b+c \end{bmatrix}$$

$$\mathbf{B}_s = \frac{1}{2} \begin{bmatrix} 1 & 1 & 1 & 1 \\ 1 & -1 & -1 & 1 \\ 1 & -1 & 1 & -1 \\ 1 & 1 & -1 & -1 \end{bmatrix}$$

$$\mathbf{S}^{-1/2} = \begin{bmatrix} 2 & 0 & 0 & 0 \\ 0 & 2 & 0 & 0 \\ 0 & 0 & 2 & 0 \\ 0 & 0 & 0 & 2 \end{bmatrix}$$

$$\lambda_2 = 2a + 2c$$

$$\lambda_3 = 2a + 2b$$

$$\lambda_4 = 2b + 2c$$

APPENDIX 4

SYMMETRY NUMBERS AND EQUILIBRIUM CONSTANTS FOR INTERCONVERTING OPTICAL AND LABEL ISOMERS

When all the species in a network of unimolecular reactions are optical or label isomers it is usually true that $\Delta H \simeq 0$ for each elementary step. In such cases equilibrium constants are controlled by entropy differences and can be calculated from symmetry numbers.

The external symmetry number of a molecule is the number of equivalent representations that can be obtained by rigid rotation. Thus for norbornane the symmetry number (σ) is 2, for norbornene it is 1, and for benzene it is 12. As one might expect the symmetry number is related to the point group of the molecule: for point groups C_n, C_{nv}, and C_{nh}, $\sigma = n$ ($n \neq \infty$). For point groups D_n, D_{nv}, and D_{nh}, $\sigma = 2n$ ($n \neq \infty$). For point groups S_n, $\sigma = n/2$.

Occasionally one has to include internal rotations in the determination of the total symmetry number. This is necessary only if a group undergoing rapid internal rotation (sufficiently rapid that all rotamers are included in a single molecular species) is involved in a reaction that moves a label among sites of different symmetry. An example would be a reaction that moves a deuterium or ^{13}C label among the *ipso*, *ortho*, *meta*, and *para* positions of a phenyl group. The total symmetry number of a molecule with a label at the *ipso* or *para* position would be twice that of a molecule with the label at the *ortho* or *meta* position (assuming that the external symmetry numbers are all equal).

For two interconverting species, A and B, the equilibrium constant K ($= [B]/[A]$) will be given by σ_A/σ_B provided that A and B are optical or label isomers (and, in the latter case, that equilibrium isotope effects are negligible). This relationship applies to single enantiomers of chiral molecules. The effective symmetry number of a racemate is half that of the corresponding single enantiomer.

The relationship between equilibrium constants and symmetry numbers allows one to generate an effective symmetrization matrix for a system of interconverting optical and label isomers, merely by determining the symmetry numbers of the species involved.

For three such species at equilibrium the fractional concentration of each can be written in terms of the symmetry numbers as follows:

$$\frac{A_1^\infty}{A_1^\infty + A_2^\infty + A_3^\infty} = \frac{1}{1 + A_2^\infty/A_1^\infty + A_3^\infty/A_1^\infty}$$

$$\frac{A_1^\infty}{A_1^\infty + A_2^\infty + A_3^\infty} = \frac{1}{\sigma_1(1/\sigma_1 + 1/\sigma_2 + 1/\sigma_3)}$$

In general, then, for a system of n optical or label isomers at equilibrium:

$$A_i^\infty = \frac{1}{\sigma_i} \; \frac{\displaystyle\sum_j^n A_j^\infty}{\displaystyle\sum_j^n (1/\sigma_j)}$$

$$= \text{constant}/\sigma_i$$

Thus the symmetrization matrix $\mathbf{S}^{-1/2}$ differs from the matrix $\sigma^{1/2}$ (whose off-diagonal elements are zero and whose diagonal elements are the square roots of the symmetry numbers of the corresponding species) only by a scalar constant. Since the symmetrization process is applied both to $\mathbf{A}$ and $\mathbf{A}^\circ$ in the integrated rate equations, the scalar constant cancels out and $\sigma^{1/2}$ can be used in place of $\mathbf{S}^{-1/2}$.

APPENDIX 5

BENSON GROUP CONTRIBUTIONS TO ΔH_f° and S°

The implied standard state for the following group values is an ideal gas at 1 atm and 25° C.

Doering [*Proc. Natl. Acad. Sci.*, **78,** 5279 (1981)] has suggested that the values for ΔH_f° of primary, secondary, and tertiary radicals calculated using the groups given here will be systematically too low by 2.6, 2.7, and 2.5 kcal/mol, respectively. These corrections apparently do not apply if the radical is involved in allylic or benzylic stabilization.

The tables are reprinted with permission from S. W. Benson, *Thermochemical Kinetics*, 2nd ed., Wiley-Interscience, New York, 1976, pp. 272–284. Copyright, John Wiley & Sons, Inc., 1976.

TABLE A.5.1. Group Values for ΔH_f°, S_{int}°, and C_{pT}°, Hydrocarbons

Group	ΔH_f° 298	S_{int}° 298	C_p°						
			300	400	500	600	800	1000	1500
C—(H)₃(C)	-10.20	30.41	6.19	7.84	9.40	10.79	13.02	14.77	17.58
C—(H)₂(C)₂	-4.93	9.42	5.50	6.95	8.25	9.35	11.07	12.34	14.25
C—(H)(C)₃	-1.90	-12.07	4.54	6.00	7.17	8.05	9.31	10.05	11.17
C—(C)₄	0.50	-35.10	4.37	6.13	7.36	8.12	8.77	8.76	8.12
Cₐ—(H)₂	6.26	27.61	5.10	6.36	7.51	8.50	10.07	11.27	13.19
Cₐ—(H)(C)	8.59	7.97	4.16	5.03	5.81	6.50	7.65	8.45	9.62
Cₐ—(C)₂	10.34	-12.70	4.10	4.61	4.99	5.26	5.80	6.08	6.36
Cₐ—(C)ₐ(H)	6.78	6.38	4.46	5.79	6.75	7.42	8.35	8.99	9.98
Cₐ—(Cₐ)(C)	8.88	-14.6	(4.40)	(5.37)	(5.93)	(6.18)	(6.50)	(6.62)	(6.72)
[Cₐ—(C_B)(H)]	6.78	6.38	4.46	5.79	6.75	7.42	8.35	8.99	9.98
Cₐ—(C_B)(C)	8.64	(-14.6)	(4.40)	(5.37)	(5.93)	(6.18)	(6.50)	(6.62)	(6.72)
[Cₐ—(Cₜ)(H)]	6.78	6.38	4.46	5.79	6.75	7.42	8.35	8.99	9.98
Cₐ—(C_B)₂	8.0								
Cₐ—(Cₐ)₂	4.6								
C—(Cₐ)(C)(H)₂	-4.76	9.80	5.12	6.86	8.32	9.49	11.22	12.48	14.36
C—(Cₐ)₂(H)₂	-4.29	(10.2)	(4.7)	(6.8)	(8.4)	(9.6)	(11.3)	(12.6)	(14.4)
C—(Cₐ)(C_B)(H)₂	-4.29	(10.2)	(4.7)	(6.8)	(8.4)	(9.6)	(11.3)	(12.6)	(14.4)
C—(Cₜ)(C)(H)₂	-4.73	10.30	4.95	6.56	7.93	9.08	10.86	12.19	14.20
C—(C_B)(C)(H)₂	-4.86	9.34	5.84	7.61	8.98	10.01	11.49	12.54	13.76

Group								
C—(C$_d$)(C)$_2$(H)	−1.48	(−11.69)	(4.16)	(5.91)	(7.34)	(8.19)	(9.46)	(10.19)	(11.28)
C—(C$_t$)(C)$_2$(H)	−1.72	(−11.19)	(3.99)	(5.61)	(6.85)	(7.78)	(9.10)	(9.90)	(11.12)
C—(C$_B$)(C)$_2$(H)	−0.98	(−12.15)	(4.88)	(6.66)	(7.90)	(8.75)	(9.73)	(10.25)	(10.68)
C—(C$_d$)(C)$_3$	1.68	(−34.72)	(3.99)	(6.04)	(7.43)	(8.26)	(8.92)	(8.96)	(8.23)
C—(C$_B$)(C)$_3$	2.81	(−35.18)	(4.37)	(6.79)	(8.09)	(8.78)	(9.19)	(8.96)	(7.63)
C$_t$—(H)	26.93	24.7	5.27	5.99	6.49	6.87	7.47	7.96	8.85
C$_t$—(C)	27.55	6.35	3.13	3.48	3.81	4.09	4.60	4.92	6.35
C$_t$—(C$_d$)	29.20	(6.43)	(2.57)	(3.54)	(3.50)	(4.92)	(5.34)	(5.50)	(5.80)
C$_t$—(C$_B$)	(29.20)	6.43	2.57	3.54	3.50	4.92	5.34	5.50	5.80
C$_B$—(H)	3.30	11.53	3.24	4.44	5.46	6.30	7.54	8.41	9.73
C$_B$—(C)	5.51	−7.69	2.67	3.14	3.68	4.15	4.96	5.44	5.98
C$_B$—(C$_d$)	5.68	−7.80	3.59	3.97	4.38	4.72	5.28	5.61	5.75
[C$_B$—(C$_t$)]	5.68	−7.80	3.59	3.97	4.38	4.72	5.28	5.61	5.75
C$_B$—(C$_B$)	4.96	−8.64	3.33	4.22	4.89	5.27	5.76	5.95	(6.05)
C$_a$	34.20	6.0	3.9	4.4	4.7	5.0	5.3	5.5	5.7
C$_{BF}$—(C$_B$)$_2$(C$_{BF}$)	4.8	−5.0	3.0	3.7	4.2	4.6	5.2	5.5	—
C$_{BF}$—(C$_B$)(C$_{BF}$)$_2$	3.7	−5.0	3.0	3.7	4.2	4.6	5.2	5.5	—
C$_{BF}$—(C$_{BF}$)$_3$	1.5	1.4	2.0	2.9	3.5	4.0	4.7	5.1	—

C$_d$ represents double-bonded C atom, C$_t$ the triple bonded C-atom, C$_B$ the C atom in a benzene ring and C$_a$ an allenic C atom. By convention group values for C—(X)(H)$_3$ will always be taken as those for C—(C)(H)$_3$ when X is any other polyvalent atom such as C$_d$, C$_t$, C$_B$, O, and S. C$_{BF}$ represents a carbon atom in a fused ring system such as naphthalene, anthracene, etc. C$_{BF}$—(C$_{BF}$)$_3$ represents the group in graphite.

TABLE A.5.2. Non-Next-Nearest Neighbor Corrections

Group	$\Delta H_f°$ 298	$S°_{int}$ 298	$C_p°$						
			300	400	500	600	800	1000	1500
Alkane *gauche* correction	0.80								
Alkene *gauche* correction	0.50								
cis-Correction	1.00[a]	[b]	-1.34	-1.09	-0.81	-0.61	-0.39	-0.26	0
ortho-Correction	0.57	-1.61	1.12	1.35	1.30	1.17	0.88	0.66	-0.05
1,5 H repulsion[c]	1.5								

[a] When one of the groups is *t*-butyl the *cis* correction = 4.00, when both are *t*-butyl, *cis* correction = ~10.00, and when there are two *cis* corrections around one double bond, the total correction is 3.00.
[b] +1.2 for but-2-ene, 0 for all other 2-enes, and -0.6 for 3-enes.
[c] These refer to repulsions between the H atoms attached to the 1,5 C atoms in such compounds as 2,2,4-tetramethyl pentane, and then only to the methyls close to each other.

Ring (σ)	ΔH_f° 298	S_{int}° 298	C_p°						
			300	400	500	600	800	1000	1500
cyclopropane (6)	27.6	32.1	−3.05	−2.53	−2.10	−1.90	−1.77	−1.62	(−1.52)
cyclopropene (2)	53.7	33.6							
cyclobutane (8)	26.2	29.8	−4.61	−3.89	−3.14	−2.64	−1.88	−1.38	−0.67
cyclobutene (2)	29.8	29.0	−2.53	−2.19	−1.89	−1.68	−1.48	−1.33	−1.22
cyclopentane (10)	6.3	27.3	−6.50	−5.5	−4.5	−3.8	−2.8	−1.93	−0.37
cyclopentene (2)	5.9	25.8	−5.98	−5.35	−4.89	−4.14	−2.93	−2.26	−1.08
cyclopentadiene (2)	6.0	28.0	−4.3						
cyclohexane (6)	0	18.8	−5.8	−4.1	−2.9	−1.3	1.1	2.2	3.3
cyclohexene (2)	1.4	21.5	−4.28	−3.04	−1.98	−1.43	−0.29	0.08	0.81
cyclohexadiene 1,3	4.8								
cyclohexadiene 1,4	0.5								
cycloheptane (1)	6.4	15.9							
cycloheptene	5.4								
cycloheptadiene, 1,3	6.6								
cycloheptatriene 1,3,5 (1)	4.7	23.7							
cyclooctane (8)	9.9	16.5							
cis-Cyclooctene	6.0								
trans-Cyclooctene	15.3								
cyclooctatriene 1,3,5	8.9								
cyclooctatetraene	17.1								
cyclononane	12.8								
cis-Cyclononene	9.9								
trans-Cyclononene	12.8								
cyclodecane	12.6								
cyclododecane	4.4								
spiropentane (4)	63.5	67.6							
cycloheptadiene	31.6								
phenylene	58.8								
cycloheptane (2,2,1)	16.2								
cyclo-(1,1,0)-butane (2)	67.0	69.2							
cyclo-(2,1,0)-pentane	55.3								
cyclo-(3,1,0)-hexane	32.7								
cyclo-(4,1,0)-heptane	28.9								
cyclo-5,1,0)-octane	29.6								
cyclo-(6,1,0)-nonane	31.1								
ethylene cyclopropane	41								

Note that in most cases the ΔH_f° correction equals ring-strain energy.

TABLE A.5.4. Oxygen-Containing Compounds

Group	ΔH_f° 298	S_{int}° 298	C_p°						
			300	400	500	600	800	1000	1500
O(H₂)	−57.8	45.1	8.0	8.4	9.2	9.9	11.2		
O(H)(C)	−37.9	29.07	4.3	4.4	4.8	5.2	6.0	6.6	
O(H)(C$_B$)^a	−37.9	29.1	4.3	4.5	4.8	5.2	6.0	6.6	
O(H)(O)	−16.3	27.85	5.2	5.8	6.3	6.7	7.2	7.5	8.2
O(H)(CO)	−58.1	24.5	3.8	5.0	5.8	6.3	7.2	7.8	
O(C)₂	−23.2	8.68	3.4	3.7	3.7	3.8	4.4	4.6	
O(C)(C$_d$)	−30.5	9.7							
O(C)(C$_B$)	−23.0								
O(C)(O)	−4.5	[9.4]	3.7	3.7	3.7	3.7	4.2	4.2	4.8
O(C)(CO)	−43.1	8.4							
O(C$_d$)₂	−33.0	10.1							
O(C$_B$)₂	−21.1								
O(C$_d$)(CO)	−45.2								
O(C$_B$)(CO)	−36.7								
O(O)(CO)	−19.0								
O(CO)₂	−46.5								
O(O)₂	[19.0]	[9.4]	[3.7]	[3.7]	[3.7]	[3.7]	[4.2]	[4.2]	[4.8]
CO(H)₂	−26.0	52.3	8.5	10.5	13.4	14.8	17.0		

$CO(H)(C)$	−29.1	34.9	7.0	7.8	8.8	9.7	11.2	12.2	
$CO(H)(C_B)$[b]	−29.1								
$CO(H)(C_d)$	−29.1								
$CO(H)(C_t)$	−29.1								
$CO(H)(CO)$	−25.3								
$CO(H)(O)$	−32.1	34.9	7.0	7.9	8.8	9.7	11.2	12.2	
$CO(C)_2$	−31.4	15.0	5.6	6.3	7.1	7.8	8.9	9.6	
$CO(C)(C_B)$	−30.9								
$CO(C_B)_2$	−25.8								
$CO(C)(O)$	−35.1	14.8	6.0	6.7	7.3	8.0	8.9	9.4	
$CO(C)(CO)$	−29.2								
$CO(C_d)(O)$[c]	−32.0								
$CO(C_B)(O)$	−36.6								
$CO(C_B)(CO)$	−26.8								
$CO(O_2)$	−29.9								
$CO(O)(CO)$	−29.3								
$C(H)_3(O)$[d]	−10.08	30.41	6.19	7.84	9.40	10.79	13.03	14.77	17.58
$C(H)_2(O)(C)$	−8.1	9.8	4.99	6.85	8.30	9.43	11.11	12.33	
$C(H)_2(O)(C_d)$	−6.5	9.7							
$C(H)_2(O)(C_B)$	−8.1								
$C(H)_2(O)(C_t)$	−6.5								
$C(H)_2(O)(CO)$									
$C(H)_2(O)_2$	−16.1								
$C(H)(O)(C)_2$	−7.2	−11.0	4.80	6.64	8.10	8.73	9.81	10.40	
$C(H)(O)_2(C)$	−16.3								

TABLE A.5.4. (Continued)

Group	ΔH_f° 298	S_{int}° 298	C_p° 300	400	500	600	800	1000	1500
C(O)(C)₃	-6.6	-33.56	4.33	6.19	7.25	7.70	8.20	8.24	
C(O)₂(C)₂^e	-18.6								
C(H)₃(CO)	-10.08	30.41	6.19	7.84	9.40	10.79	13.02	14.77	17.58
C(H)₂(CO)(C)	-5.2	9.6	6.2	7.7	8.7	9.5	11.1	12.2	
C(H)₂(CO)(Cd)	-3.8								
C(H)₂(CO)(CB)	-5.4								
C(H)₂(CO)(Ct)	-5.4								
C(H)₂(CO)₂	-7.6								
C(H)(CO)(C)₂^f	-1.7	-12.0							
C(CO)(C)₃	1.4								
Cd(O)(H)^g	8.6	8.0	4.2	5.0	5.8	6.5	7.6	8.4	9.6
Cd(O)(C)^h	10.3								
Cd(O)(Cd)^i	8.9								
Cd(O)(CO)	11.6								
Cd(H)(CO)	5.0								
Cd(CO)(CO)	7.5								
CB(O)	-0.9	-10.2	3.9	5.3	6.2	6.6	6.9	6.9	
CB(CO)	3.7								

^{a} O(H)(Cd)≡O(H)(CB)≡O(H)(Ct)≡O(H)(C), assigned.
^{b} CO(t.._CB)≡CO(H)(C), assigned.
^{c} CO(C)(Cd)≡CO(CB)(O), assigned.
^{d} C(H)₃(O)≡C(H)₃(C), assigned.
^{e} C(H)₃(CO)≡C(H)₃(C), assigned.
^{f} C(H)(CO)(C₂), estimated.
^{g} Cd(H)(O)≡Cd(H)(C), assigned.
^{h} Cd(O)(C)≡Cd(C)₂
^{i} Cd(O)(Cd)≡Cd(C)(Cd).

228

TABLE A.5.5. Non-Nearest-Neighbor and Ring Corrections

Strain	ΔH°_f 298	S°_{int} 298	C°_p						
			300	400	500	600	800	1000	1500
Ether-oxygen *gauche*	0.5								
Di-tertiary ethers	7.8								
Oxygen *gauche*	0								
Oxygen *ortho*	0								
(3-membered ring with O)	−26.9	30.5	−2.0	−2.8	−3.0	−2.6	−2.3	−2.3	
(4-membered ring with O)	25.7	27.7	−4.6	−5.0	−4.2	−3.5	−2.6	+0.2	
(5-membered ring with O)	5.9								
(6-membered ring with O)	0.5								
(6-membered ring with two O)	0.2								
(6-membered ring with two O, para)	3.3								

TABLE A.5.5. (Continued)

Strain	ΔH°_f 298	S°_{int} 298	C°_p						
			300	400	500	600	800	1000	1500
	6.6								
	4.7								
	6.0								
	−5.8								
	1.2								
	4.5								
	0.8								
	3.6								

Ring Correction	ΔH°_f 298	Ring Correction	ΔH°_f 298
(benzodioxole structure)	16.6	(cyclobutanone structure)	22.6
(benzodioxane structure)	2.0	Cyclopentanone	5.2
		Cyclohexanone	2.2
		Cycloheptanone	2.3
(xanthene structure)	2.3	Cyclooctanone	1.5
		Cyclononanone	4.7
		Cyclodecanone	3.6
(octahydrodibenzofuran structure)	11.4	Cycloundecanone	4.4
(decalone structure)	cis 15.3 trans 20.9	Cyclododecanone	3.0
(β-propiolactone structure)	23.9	Cyclo(C_{15})anone	2.1
(diketene structure)	22.0	Cyclo(C_{17})anone	1.1

TABLE A.5.6. Group Contributions to $C_{p,T}^\circ$, S°, and ΔH_f° at 25°C and 1 atom for Nitrogen-Containing Compounds

Group	ΔH_f° 298	S_{int}° 298	C_p°						
			300	400	500	600	800	1000	1500
C—(N)(H)$_3$	−10.08	30.41	6.19	7.84	9.40	10.79	13.02	14.77	17.58
C—(N)(C)(H)$_2$	−6.6	9.8^a	5.25^a	6.90^a	8.28^a	9.39^a	11.09^a	12.34^a	
C—(N)(C)$_2$(H)	−5.2	−11.7^a	4.67^a	6.32^a	7.64^a	8.39^a	9.56^a	10.23^a	
C—(N)(C)$_3$	−3.2	−34.1^a	4.35^a	6.16^a	7.31^a	7.91^a	8.49^a	8.50^a	
N—(C)(H)$_2$	4.8	29.71	5.72	6.51	7.32	8.07	9.41	10.47	12.28
N—(C)$_2$(H)	15.4	8.94	4.20	5.21	6.13	6.83	7.90	8.65	9.55
N—(C)$_3$	24.4	−13.46	3.48	4.56	5.43	5.97	6.56	6.67	6.50
N—(N)(H)$_2$	11.4	29.13	6.10	7.38	8.43	9.27	10.54	11.52	13.19
N—(N)(C)(H)	20.9	9.61	4.82	5.8	6.5	7.0	7.8	8.3	9.0
N—(N)(C)$_2$	29.2	−13.80							
N—(N)(C$_B$)(H)	22.1								
N$_I$—(H)	16.3								
N$_I$(C)	21.3								
N$_I$(C$_B$)^b	16.7								
N$_A$—(H)	25.1	26.8	4.38	4.89	5.44	5.94	6.77	7.42	8.44
N$_A$—(C)	27								
N—(C$_d$)(C)(H)	15.4								
N—(C$_d$)(C)(N)	30								
N—(N$_I$)(C)(H)	21								
N—(C$_d$)(H)$_2$	4.8								
N—(C$_d$)(C)$_2$	24.4								
N—(C$_d$)(H)(N)	21.5								

Group									
N—(C_B)(H)_2	4.8	29.71	5.72	6.51	7.32	8.07	9.41	10.47	12.28
N—(C_B)(C)(H)	14.9								
N—(C_B)(C)_2	26.2								
N—(C_B)_2(H)	16.3								
N_A—(N)	23.0								
C_B—(N)	-0.5	-9.69	3.95	5.21	5.94	6.32	6.53	6.56	
CO—(N)(H)	-29.6	34.93	7.03	7.87	8.82	9.68	11.16	12.20	
CO—(N)(C)	-32.8	16.2	5.37	6.17	7.07	7.66	9.62	11.19	
N—(CO)(H)_2	-14.9	24.69	4.07	5.74	7.13	8.29	9.96	11.22	
N—(CO)(C)(H)	-4.4	3.9[a]							
N—(CO)(C)_2	+0.4								
N—(CO)(C_B)(H)	-18.5								
N—(CO)_2(H)	-5.9								
N—(CO)_2(C)	-0.5								
N—(CO)_2(C_B)									
C—(N_A)(C)(H)_2	-6.0								
C—(N_A)(C)_2(H)	-3.4								
C—(N_A)(C)_3	-3.0								
C—(CN)(C)(H)_2	22.5	40.20	11.10	13.40	15.50	17.20	19.7	21.30	
C—(CN)(C)_2(H)	25.8	19.80	11.00	12.70	14.10	15.40	17.30	18.60	
C—(CN)(C)_3	29.0	-2.80							
C—(CN)_2(C)_2		28.40							
C_d—(CN)(H)	37.4	36.58	9.80	11.70	13.30	14.50	16.30	17.30	
C_d—(CN)_2	34.1								
C_d—(NO_2)(H)	35.8	44.4	12.3	15.1	17.4	19.2	21.6	23.2	25.3
C_B—(CN)	63.8	20.50	9.8	11.2	12.3	13.1	14.2	14.9	
C_t—(CN)	63.8	35.40	10.30	11.30	12.10	12.70	13.60	14.30	15.30
C—(NCO)	-10.2	48.9	15.4						
C—(NO_2)(C)(H)_2	-15.1	48.4[a]							

TABLE A.5.6. (*Continued*)

Group	ΔH_f° 298	S_{int}° 298	C_p° 300	400	500	600	800	1000	1500
C—(NO₂)(C)₂(H)	−15.8	26.9[a]							
C—(NO₂)(C)₃		3.9[a]							
C—(NO₂)₂(C)(H)	−14.9								
O—(NO)(C)	−5.9	41.9	9.10	10.30	11.2	12.0	13.3	13.9	14.5
O—(NO₂)(C)	−19.4	48.50							
O—(C)(CN)	2.0	39.5	10.0						
O—(C_d)(CN)	7.5	43.1	13.0						
O—(C_B)(CN)	7.0	29.2	8.3						

Corrections To Be Applied to Ring-Compound Estimates

Group	ΔH_f° 298	S_{int}° 298	C_p° 300	400	500	600	800	1000	1500
Ethyleneimine	27.7	31.6[a]							
Azetidine	26.2[a]	29.3[a]							
Pyrrolidine	6.8	26.7	−6.17	−5.58	−4.80	−4.00	−2.87	−2.17	

Piperidine   1.0

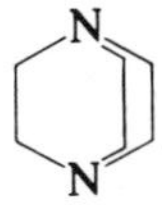

3.4

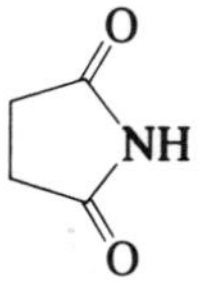

8.5

---

[a] Estimates by Benson.
[b] For *ortho* or *para* substitution in pyridine add −1.5 kcal/mol per group.
N_I stands for imino N atom; N_A represents azo N atom.

$C-(N_I)(C)(H)_2 \equiv C-(N)(C)(H)_2$; assigned
$C-(N_I)(C)_2(H) \equiv C-(N)(C)_2(H)$; $C_d-(N_I)(H) = C_d-(C_d)(H)$

235

TABLE A.5.7. Halogen-Containing Compounds. Group Contribution to ΔH_{f298}°, S_{298}°, and C_{pT}°, Ideal Gas at 1 atm

Group	ΔH_f° 298	S_{int}° 298	C_p°					
			300	400	500	600	800	1000
C—(F)₃(C)	−158.4	42.5	12.7	15.0	16.4	17.9	19.3	20.0
C—(F)₂(H)(C)	(−102.3)	39.1	9.9	12.0		15.1		
C—(F)(H)₂(C)	−51.5	35.4	8.1	10.0	12.0	13.0	15.2	16.6
C—(F)₂(C)₂	−97.0	17.8	9.9	11.8	13.5			
C—(F)(H)(C)₂	−49.0	(14.0)						
C—(F)(C)₃	−48.5							
C—(F)₂(Cl)(C)	−106.3	40.5	13.7	16.1	17.5			
C—(Cl)₃(C)	−20.7	50.4	16.3	18.0	19.1	19.8	20.6	21.0
C—(Cl)₂(H)(C)	(−18.9)	43.7	12.1	14.0	15.4	16.5	17.9	18.7
C—(Cl)(H)₂(C)	−16.5	37.8	8.9	10.7	12.3	13.4	15.3	16.7
C—(Cl)₂(C)₂	−22.0	22.4	12.2					
C—(Cl)(H)(C)₂	−14.8	17.6	9.0	9.9	10.5	11.2		
C—(Cl)(C)₃	−12.8	5.4	9.3	10.5	11.0	11.3		
C—(Br)₃(C)		55.7	16.7	18.0	18.8	19.4	19.9	20.3
C—(Br)(H)₂(C)	−5.4	40.8	9.1	11.0	12.6	13.7	15.5	16.8
C—(Br)(H)(C)₂	−3.4	−2.0	9.3	11.0				
C—(Br)(C)₃	−0.4		9.2					
C—(I)(H₂)(C)	8.0	43.0	9.2	11.0	12.9	13.9	15.8	17.2
C—(I)(H)(C)₂	10.5	21.3	9.7	10.9	12.2	13.0	14.2	14.8
C—(I)(C)₃	13.0	0.0						
C—(I)₂(C)(H)	(26.0)	(54.6)	(12.2)	—	(16.4)	(17.0)		
C—(Cl)(Br)(H)(C)		45.7	12.4	14.0	15.6	16.3	17.9	19.0
N—(F)₂(C)	−7.8							
C—(Cl)(C)(O)(H)	−21.6	15.9	(9.0)	(9.9)	(10.5)	(11.2)		

C—(I)(O)(H)₂	3.8	40.7						
Cd—(C)(Cl)	−2.1	15.0						
Cd—(F)₂	−77.5	37.3	9.7	11.0	12.0	12.7	13.8	14.5
Cd—(Cl)₂	−1.8	42.1	11.4	12.5	13.3	13.9	14.6	15.0
Cd—(Br)₂		47.6	12.3	13.2	13.9	14.3	14.9	15.2
Cd—(F)(Cl)		39.8	10.3	11.7	12.6	13.3	14.2	14.7
Cd—(F)(Br)		42.5	10.8	12.0	12.8	13.5	14.3	14.7
Cd—(Cl)(Br)		45.1	12.1	12.7	13.5	14.1	14.7	14.7
Cd—(F)(H)	−37.6	32.8	6.8	8.4	9.5	10.5	11.8	12.7
Cd—(Cl)(H)	−1.2	35.4	7.9	9.2	10.3	11.2	12.3	13.1
Cd—(Br)(H)	11.0	38.3	8.1	9.5	10.6	11.4	12.4	13.2
Cd—(I)(H)	24.5	40.5	8.8	10.0	10.9	11.6	12.6	13.3
Ct—(Cl)		33.4	7.9	8.4	8.7	9.0	9.4	9.6
Ct—(Br)		36.1	8.3	8.7	9.0	9.2	9.5	9.7
Ct—(I)		37.9	8.4	8.8	9.1	9.3	9.6	9.8
CB—(F)	−42.8	16.1	6.3	7.6	8.5	9.1	9.8	10.2
CB—(Cl)	−3.8	18.9	7.4	8.4	9.2	9.7	10.2	10.4
CB—(Br)	10.7	21.6	7.8	8.7	9.4	9.9	10.3	10.5
CB—(I)	24.0	23.7	8.0	8.9	9.6	9.9	10.3	10.5
C—(CB)(F)₃	−162.7	42.8	12.5	15.3	17.2	18.5	20.1	21.0
C—(CB)(Br)(H)₂	−5.1							
C—(CB)(I)(H)₂	8.4							

Corrections for Non-Next-Nearest Neighbors

ortho (F)(F)	5.0	0	0	0	0	0	0	0
ortho (Cl)(Cl)	2.2							
ortho (alk) (halogen)ᵃ	0.6							
cis (halogen) (halogen)	−0.3							
cis (halogen)(alk)	−0.8							

ᵃHalogen = Cl, Br, I only.
The *gauche* correction = 1.0 kcal for Cl, Br, I; none for X— Me and none for F—halogen.

237

TABLE A.5.8. Sulfur-Containing Compounds. Group Contributions to $\Delta H_f^\circ 298$, $S^\circ 298$, and C_{pT}°

Group	ΔH_f° 298	S_{int}° 298	C_p° 300	400	500	600	800	1000
C—(H)₃(S)ᵃ	−10.08	30.41	6.19	7.84	9.40	10.79	13.02	14.77
C—(C)(H)₂(S)	−5.65	9.88	5.38	7.08	8.60	9.97	12.26	14.15
C—(C)₂(H)(S)	−2.64	−11.32	4.85	6.51	7.78	8.69	9.90	10.57
C—(C)₃(S)	−0.55	−34.41	4.57	6.27	7.45	8.15	8.72	8.10
C—(C_B)(H)₂(S)	−4.73							
C—(C_d)(H)₂(S)	−6.45							
C—(H)₂(S)₂	−6.0 ± 3							
C_B—(S)ᵇ	−1.8	10.20	3.90	5.30	6.20	6.60	6.90	6.90
C_d—(H)(S)ᶜ	8.56	8.0	4.16	5.03	5.81	6.50	7.65	8.45
C_d—(C)(S)	10.93	−12.41	3.50	3.57	3.83	4.09	4.41	5.00
S—(C)(H)	4.62	32.73	5.86	6.20	6.51	6.78	7.30	7.71
S—(C_B)(H)	11.96	12.66	5.12	5.26	5.57	6.03	6.99	7.84
S—(C)₂	11.51	13.15	4.99	4.96	5.02	5.07	5.41	5.73
S—(C)(C_d)	9.97							
S—(C_d)₂	−4.54	16.48	4.79	5.58	5.53	6.29	7.94	9.73
S—(C_B)(C)	19.16							
S—(C_B)₂	25.90							
S—(S)(C)	7.05	12.37	5.23	5.42	5.51	5.51	5.38	5.12
S—(S)(C_B)	14.5							
S—(S)₂	3.01	13.4	4.7	5.0	5.1	5.2	5.3	5.4
C—(SO)(H)₃ᵈ	−10.08	30.41	6.19	7.84	9.40	10.79	13.02	14.77
C—(C)(SO)(H)₂	−7.72							
C—(C)₃(SO)	−3.05							
C—(C_d)(SO)(H)₂	−7.35							

Group								
C_B—(SO)[e]	2.3							
SO—(C)$_2$	−14.41	18.10	8.88	10.03	10.50	10.79	10.98	11.17
SO—(C$_B$)$_2$	−12.0		6.19	7.84	9.40	10.79	13.02	14.77
C—(SO$_2$)(H)$_3$[f]	−10.08	30.41						
C—(C)(SO$_2$)(H)$_2$	−7.68							
C—(C)$_2$(SO$_2$)(H)	−2.62							
C—(C)$_3$(SO$_2$)	−0.61							
C—(C$_d$)(SO$_2$)(H)$_2$	−7.14							
C—(C$_B$)(SO$_2$)(H)$_2$	−5.54							
C$_B$—(SO$_2$)[g]	2.3							
C$_d$—(H)(SO$_2$)	12.5							
C$_d$—(C)(SO$_2$)	14.5							
SO$_2$—(C$_d$)(C$_B$)	−68.6							
SO$_2$—(C$_d$)$_2$	−73.6							
SO$_2$—(C)$_2$	−69.74	20.90	11.52					
SO$_2$—(C)(C$_B$)	−72.29							
SO$_2$—(C$_B$)$_2$	−68.58							
SO$_2$—(SO$_2$)(C$_B$)	−76.25							
CO—(S)(C)[h]	−31.56	15.43	5.59	6.32	7.09	7.76	8.89	9.61
S—(H)(CO)	−1.41	31.20	7.63	8.09	8.12	8.17	8.50	8.24
C—(S)(F)$_3$	−31.56	38.9	5.59	6.32	7.09	7.76	8.89	9.61
CS—(N)$_2$[i]	12.78	29.19	6.07	7.28	8.18	8.91	10.09	10.98
N—(CS)(H)$_2$	−4.90							
S—(S)(N)[j]	29.9							
N—(S)(C)$_2$	−31.56							
SO—(N)$_2$[k]	16.0							
N—(SO)(C)$_2$	−31.56							
SO$_2$—(N)$_2$[l]	−31.56							
N—(SO$_2$)(C)$_2$	−20.4							

Corrections To Be Applied to Organosulfur Ring Compounds

Structure									
S (3-membered ring)	(2)	17.7	29.5	−2.9	−2.6	−2.7	−3.0	−4.3	−5.8
S (4-membered ring)	(2)	19.4	27.2	−4.6	−4.2	−3.9	−3.9	−4.6	−5.7
S (5-membered ring)	(2)	1.7	23.6	−4.9	−4.7	−3.7	−3.7	−4.4	−5.6
S (6-membered ring)	(1)	0	16.1	−6.2	−4.3	−2.8	−0.7	−0.9	−1.3
S (7-membered ring)	(1)	3.9							
S (5-membered ring, unsaturated)	(2)	5.0							
S (5-membered ring, unsaturated)	(1)	5.0							

| | (2) | 5.7 | | | | | |
| 23.6 | −4.9 | −4.7 | −3.7 | −3.7 | −4.4 | −5.6 |

| | (2) | 1.7 |

[a] $C-(S)(H)_3 \equiv C-(C)(H)_3$, assigned.
[b] $C_B-(S) \equiv C_B-(O)$, assigned.
[c] $C_d-(S)(H) \equiv C_d-(O)(H)$, assigned.
[d] $C-(SO)(H)_3 \equiv C-(CO)(H)_3$, assigned.
[e] $C_B-(SO) \equiv C_B-(CO)$, assigned.
[f] $C-(SO_2)(H)_3 \equiv C-(SO)(H)_3$.
[g] $C_B-(SO_2) \equiv C_B-(CO)$, assigned.
[h] $CO-(S)(C) \equiv CO-(C)_2$, assigned.
[i] $CS-(N)_2 \equiv CO-(C)_2$, assigned.
[j] $S-(S)(N) \equiv O-(O)(C)$, assigned.
[k] $SO-(N)_2 \equiv CO-(C)_2$, assigned.
[l] $SO_2-(N)_2 \equiv SO-(N)_2$, assigned.

[m] Assume ring corrections for and are the same.

[n] Assume ring corrections for and are the same.

241

TABLE A.5.9. Free-Radical Group Additivities[a]

Radical	ΔH_f°	S°	C_p° 300	400	500	600	800	1000	1500
[·C—(C)(H)₂]	35.82	30.7	5.99	7.24	8.29	9.13	10.44	11.47	13.14
[·C—(C)₂(H)]	37.45	10.74	5.16	6.11	6.82	7.37	8.26	8.84	9.71
[·C—(C)₃]	38.00	-10.77	4.06	4.92	5.42	5.75	6.27	6.35	6.53
[C—(C·)(H)₃]	-10.08	30.41	6.19	7.84	9.40	10.79	13.02	14.77	17.58
[C—(C·)(C)(H)₂]	-4.95	9.42	5.50	6.95	8.25	9.35	11.07	12.34	14.25
[C—(C·)(C)₂(H)]	-1.90	-12.07	4.54	6.00	7.17	8.05	9.31	10.05	11.17
[C—(C·)(C)₃]	1.50	-35.10	4.37	6.13	7.36	8.12	8.77	8.76	8.12
[C—(O·)(C)(H)₂]	6.1	36.4	7.9	9.8	10.8	12.8	15.0	16.4	—
[C—(O·)(C)₂(H)]	7.8	14.7	7.7	9.5	10.6	12.1	13.7	14.5	—
[C—(O·)(C)₃]	8.6	-7.5	7.2	9.1	9.8	11.1	12.1	12.3	—
[C—(S·)(C)(H)₂]	32.4	39.0	9.0	10.6	12.4	13.6	15.8	17.4	—
[C—(S·)(C)₂(H)]	35.5	17.8	8.5	10.0	11.6	12.3	13.8	14.6	—
[C—(S·)(C)₃]	37.5	-5.3	8.2	9.8	11.3	11.8	12.2	12.3	—
[·C—(H)₂C_d)]	23.2	27.65	5.39	7.14	8.49	9.43	11.04	12.17	14.04
[·C—(H)(C)(C_d)]	25.5	7.02	4.58	6.12	7.19	8.00	9.11	9.78	10.72
[·C—(C)₂(C_d)]	24.8	-15.00	4.00	4.73	5.64	6.09	6.82	7.04	7.54
[C_d—(C·)(H)]	8.59	7.97	4.16	5.03	5.81	6.50	7.65	8.45	9.62
[C_d—(C·)(C)]	10.34	-12.30	4.10	4.71	5.09	5.36	5.90	6.18	6.40
[C·—(C_B)(H)₂]	23.0	26.85	6.49	7.84	9.10	9.98	11.34	12.42	14.14
[·C—(C_B)(C)(H)]	24.7	6.36	5.30	6.87	7.85	8.52	9.38	9.84	10.12
[·C—(C_B)(C)₂]	25.5	-15.46	4.72	5.48	6.20	6.65	7.09	7.10	6.94

242

[C$_B$—C·]	5.51	−7.69	2.67	3.14	3.68	4.15	4.96	5.44	5.98
C—(·CO)(H)$_3$	−5.4	66.6	12.74	14.63	16.47	18.17	21.14	23.27	—
C—(·CO)(C)(H)$_2$	−0.3	45.8	12.7	14.5	15.8	16.8	19.2	20.7	—
C—(·CO)(C)$_2$(H)	2.6	(23.7)	(11.5)	(12.8)	(14.3)	(15.5)	(17.4)	(18.5)	7.74
·N—(H)(C)	(55.3)	30.23	5.38	5.67	5.89	6.09	6.60	6.97	4.91
·N—(C)$_2$	(58.4)	10.24	3.72	4.13	4.38	4.53	4.86	4.95	—
C—(·N)(C)(H)$_2$	−6.6	9.8	5.25	6.90	8.28	9.39	11.09	12.34	—
C—(·N)(C)$_2$(H)	−5.2	−11.7	4.67	6.32	7.64	8.39	9.56	10.23	—
C—(·N)(C)$_3$	(−3.2)	34.1	4.35	6.16	7.31	7.91	8.49	8.50	—
·C—(H)$_2$(CN)	(58.2)	58.5	10.66	12.82	14.48	15.89	18.08	19.80	8.6
·C—(H)(C)(CN)	(56.8)	40.0	9.1	11.4	13.1	14.4	16.3	17.4	(5.9)
·C—(C)$_2$(CN)	(56.1)	19.6	8.8	10.4	11.3	12.3	13.7	14.5	—
·N—(H)(C$_B$)	38.0	27.3	4.6	5.4	6.0	6.4	7.2	7.7	—
·N—(C)(C$_B$)	42.7	(6.5)	(3.9)	(4.2)	(4.7)	(5.0)	(5.6)	(5.8)	—
C$_B$—N·	−0.5	−9.69	3.95	5.21	5.94	6.32	6.53	6.56	—
C—(CO$_2$·)(H)$_3$	−47.5	71.4	14.4	17.8	20.4	23.1	27.1	29.6	—
C—(CO$_2$·)(H)$_2$(C)	−41.9	49.8	15.5	18.5	20.3	22.3	27.5	27.2	—
C—(CO$_2$·)(H)(C)$_2$	−39.0	−12.1	(4.5)	(6.0)	(7.2)	(8.0)	(9.3)	(10.1)	(11.2)
C—(N$_A$)(H)$_3$	−10.08	30.41	6.19	7.84	9.40	10.79	13.02	14.77	17.58
C—(N$_A$)(C)(H)$_2$	−5.5	9.42	5.50	6.95	8.25	9.35	11.07	12.34	14.25
C—(N$_A$)(C)$_2$(H)	−3.3	−12.07	4.54	6.00	7.17	8.05	9.31	10.05	11.17
C—(N$_A$)(C)$_3$	−1.9	−35.10	4.37	6.13	7.36	8.12	8.77	8.76	8.12
[N$_A$—C]	32.5	8.0	4.0	4.4	4.7	4.8	5.1	5.3	5.2
[N$_A$—(N$_A$·)(C)]	74.2	36.1	7.8	8.2	8.4	8.6	8.9	9.0	9.0

^{a}Values in parentheses are best guesses.

INDEX